Imported Foods
Microbiological Issues and Challenges

Emerging Issues in Food Safety

SERIES EDITOR, Michael P. Doyle

Microbiology of Fresh Produce
Edited by Karl R. Matthews

Microbial Source Tracking
Edited by Jorge W. Santo Domingo and Michael J. Sadowsky

Microbial Risk Analysis of Foods
Edited by Donald W. Schaffner

Enterobacter sakazakii
Edited by Jeffrey M. Farber and Stephen J. Forsythe

Food-Borne Viruses: Progress and Challenges
Edited by Marion P. G. Koopmans, Dean O. Cliver, and Albert Bosch

Imported Foods: Microbiological Issues and Challenges
Edited by Michael P. Doyle and Marilyn C. Erickson

Imported Foods
Microbiological Issues and Challenges

EDITED BY

Michael P. Doyle
Center for Food Safety
University of Georgia
Griffin, Georgia

AND

Marilyn C. Erickson
Center for Food Safety
University of Georgia
Griffin, Georgia

ASM
PRESS

WASHINGTON, DC

Address editorial correspondence to ASM Press, 1752 N St., N.W.,
Washington, DC 20036-2904, USA

Send orders to ASM Press, P.O. Box 605, Herndon, VA 20172, USA
Phone: 800-546-2416; 703-661-1593
Fax: 703-661-1501
E-mail: books@asmusa.org
Online: http://estore.asm.org

Library of Congress Cataloging-in-Publication Data

Imported foods: microbiological issues and challenges / edited by Michael
P. Doyle and Marilyn C. Erickson.
    p. ; cm.—(Emerging issues in food safety)
    Includes bibliographical references and index.
    ISBN-13: 978-1-55581-413-7 (hardcover: alk. paper)
    ISBN-10: 1-55581-413-1 (hardcover: alk. paper) 1. Food—
    Microbiology—United States.   2. Food contamination—United
    States—Prevention.   3. Foodborne diseases—United States—
    Prevention.   4. Food—Safety measures—United States.   5. Food
    supply—Health aspects—United States.   6. Food industry and
    trade—Health aspects—United States.   I. Doyle, Michael P., 1949–
    II. Erickson, Marilyn C.   III. Series.
        [DNLM: 1. Food Contamination—United States.
    2. Commerce—United States.   3. Food Microbiology—United States.
    4. Food Supply—United States.   5. Health Policy—United States.
    WA 701 I346 2008]

QR115.I38   2008
363.19′26—dc22   2008008231

10 9 8 7 6 5 4 3 2 1

Cover illustration: Naturally occurring biofilm developed over 3 weeks
on a stainless steel chip. The chip was glued to a chilling unit in a meat
processing environment. The biofilm was stained with acridine orange
and visualized with epifluorescent microscopy. © Amy Lee Wong,
author. Licensed for use, ASM MicrobeLibrary
(http://www.microbelibrary.org).

Contents

Contributors

JULIE A. CASWELL
Department of Resource Economics, 215 Stockbridge Hall, University of
Massachusetts, 80 Campus Center Way, Amherst, MA 01003–9246

FENGXIA DONG
Center for Agricultural and Rural Development, Iowa State University, Ames, IA
50011–1070

MICHAEL P. DOYLE
Center for Food Safety and Department of Food Science and Technology,
University of Georgia, 1109 Experiment Street, Griffin, GA 30223–1797

MARILYN C. ERICKSON
Center for Food Safety and Department of Food Science and Technology,
University of Georgia, 1109 Experiment Street, Griffin, GA 30223–1797

W. J. FLORKOWSKI
Department of Agricultural and Applied Economics, University of Georgia,
Griffin, GA 30223

NEAL D. FORTIN
Institute for Food Laws and Regulations, Michigan State University, East Lansing,
MI 48824

CHARLES P. GERBA
Department of Soil, Water and Environmental Science, University of Arizona,
Tucson, AZ 85721

HELEN H. JENSEN
Department of Economics, Center for Agricultural and Rural Development, Iowa State University, Ames, IA 50011–1070

MARTYN KIRK
OzFoodNet, Office of Health Protection, Department of Health and Ageing, GPO Box 9848, MDP 14, Canberra 2601, Australian Capital Territory, Australia

SARAH J. O'BRIEN
University of Manchester, Clinical Sciences Building, Hope Hospital, Stott Lane, Salford M6 8HD, United Kingdom

CHRISTOPHER A. SCOTT
Department of Geography and Regional Development, and Udall Center for Studies in Public Policy, University of Arizona, Tucson, AZ 85721

ROBERT V. TAUXE
Division of Foodborne, Bacterial and Mycotic Diseases, National Center for Zoonotic, Vectorborne and Enteric Diseases, Centers for Disease Control and Prevention, 1600 Clifton Road, Mailstop C-09, Atlanta, GA 30333

EWEN TODD
Food Safety Policy Center, 328 Communications Arts and Sciences Building, Michigan State University, East Lansing, MI 48824

GARNETT E. WOOD
Office of Food Safety, Center for Food Safety and Applied Nutrition, Food and Drug Administration, College Park, MD 20740

SHAOHUA ZHAO
Division of Animal and Food Microbiology, Office of Research, Center for Veterinary Medicine, U.S. Food and Drug Administration, Laurel, MD 20708

Series Editor's Foreword

The microbiological safety of foods has become a highly visible international issue of concern, with many outbreaks of food-borne illness being reported daily around the globe. In the United States alone, more than 1,200 food-borne outbreaks are reported annually, and more than 70 million cases of food-borne illness are estimated. The *Emerging Issues in Food Safety* series was conceived by ASM Press in 2006 to provide in-depth, state of the science information regarding current topics in the microbiological food safety arena.

Since the inception of the series, six volumes have been published addressing the issues of the day. Food attribution studies of cases associated with recent food-borne illness outbreaks have revealed vegetables and noroviruses as the leading vehicle and pathogen, respectively, responsible for these outbreaks. *Microbiology of Fresh Produce*, edited by Karl Matthews, and the treatise *Food-Borne Viruses: Progress and Challenges*, edited by Marion Koopmans, Dean Cliver, and Albert Bosch, provide timely, insightful perspectives on both of these topics.

Public health professionals, largely because of major advances in surveillance systems, have unearthed outbreaks likely to have gone unnoticed in the past. Jorge Santo Domingo and Michael Sadowsky have, in the volume *Microbial Source Tracking*, provided an intriguing look into the future of novel approaches to trace pathogens to their points of origin. Coupled with better surveillance systems for detecting food-borne outbreaks, this tracking technology will enable food producers and processors to better understand sources of contamination and thereby take corrective actions to prevent pathogen contamination.

New food-borne pathogens continue to emerge and sometimes reemerge, including *Enterobacter sakazakii*, which has recently become a major nemesis of the infant formula industry. Two internationally recognized leaders in studying the food-associated aspects of this pathogen, Jeffrey Farber and Stephen Forsythe, have edited the book *Enterobacter sakazakii*, which provides a comprehensive treatise on the food safety aspects of this microorganism.

On the heels of food safety issues associated with emerging food-borne pathogens are food safety concerns linked to emerging markets that provide foods and food ingredients. Many of the emerging food sources are developing countries that do not employ the same level of sanitary practices in food production and processing that developed countries do. This can result in unintended and unacceptable pathogen contamination of foods imported by countries with high standards of sanitary practices for food production and preparation. Marilyn Erickson and I have organized *Imported Foods: Microbiological Issues and Challenges* to address many of the food safety issues facing the United States and many other countries that import foods from countries having less than adequate food production and preparation practices.

With the emergence of newly recognized food-borne pathogens, major sources of foods produced under unsanitary conditions, changes in eating behavior involving more consumption of higher-risk fresh, uncooked ready-to-eat food (such as fresh vegetables), and better methods for detecting and tracking food-borne disease outbreaks, microbiological food safety issues will undoubtedly be of major international concern for many years to come. One of the greatest challenges in addressing these issues is determining how best to regulate them. Most countries have limited, and often minimal, resources to ensure the safety of foods. Hence, international efforts are being directed to developing science-based decision-making tools for identifying the most effective use of resources to minimize food-borne illnesses. The concept of microbial risk analysis has been conceived and widely embraced to address this need. Donald Schaffner has brought together a cast of internationally recognized experts to draft *Microbial Risk Analysis of Foods*, which elucidates how risk analysis of food-borne agents of microbial origin can be used to provide greater public health protection to the food supply.

Collectively, these six contributions cover the gamut of today's microbiological food safety issues and provide cutting edge information and insights that cannot be found elsewhere. I personally find them to be a treasure trove of intelligence, and they are valuable resources for food safety professionals who are at the cutting edge of food safety.

MICHAEL P. DOYLE, Series Editor
Emerging Issues in Food Safety

Preface

According to USDA-ERS data, in 2004 the United States began to import more food than it exported based on dollar value, and this differential continues to grow. Currently about 15% of foods consumed in the United States are imported, but differences in import percentages exist for commodity types. For example, in 2005, about 80% of fresh and frozen fish and shellfish, 44% of fresh fruits, 43% of tree nuts, and 16% of fresh vegetables consumed in the United States were imported. Unfortunately, sanitation practices for food production and preparation are not universally equivalent throughout the world. Many developing countries, including those that provide food to the United States, have high incidences of infectious diseases and large portions of their populations are asymptomatic carriers of food-borne pathogens like norovirus, *Campylobacter*, and parasites. These harmful microbes frequently contaminate human sewage that is used untreated to fertilize land for growing produce and ponds for growing fish and shrimp in many East Asian countries like China. Current estimates indicate at least two-thirds of the world production of farmed fish is grown in ponds fertilized with animal manure and/or human sewage. In Mexico City with its population of more than 25 million, less than 10% of the city's wastewater sewage is treated, with the remainder being sent into rivers that irrigate farmland to the north.

In addition to microbiological food safety concerns, there are many chemical contaminants associated with foods produced in developing countries. For example, in China, where farmers largely grow crops or fish in one-acre parcels, excessive or inappropriate use of pesticides, antibiotics, and veterinary drugs frequently occurs to enable maximum productivity. Consequently, antibiotics like chloramphenicol that are not allowed for crop or aquaculture purposes in the United States have been detected in foods from China.

Food distributors and processors within the United States are the first line of defense to ensure safety of the foods they make available to domestic consumers. The federal government, on the other hand, is responsible for ensuring that those companies marketing foods and food ingredients from foreign sources are providing safe foods. In particular, the U.S. Food and Drug Administration (FDA) has oversight of about 80% of the U.S. food supply but visually inspects only about 1% of about 9 million food shipments annually while less than 0.5% of imported foods are sampled and tested.

FDA testing of imported foods continues to identify tainted products. For example, in August 2007 the FDA reported 187, 173, and 160 food refusal actions from China, India, and Mexico, respectively. Examples of rejected foods from China include chives (*Salmonella*), shrimp (nitrofuran), peppercorns (*Salmonella*), pear juice concentrate (pesticides), catfish (veterinary drugs), diced green bell peppers (pesticides), IQF butterfly shrimp (veterinary drugs and nitrofuran), aniseed powder (*Salmonella*), wheat gluten (poisons, filth), and soy protein (poisonous, unsafe additives), and frozen soybeans (pesticides). *Salmonella* and pesticide contamination of spices and seasonings were common causes of refusals from India. Pesticides in produce were also a major reason for rejections of food from Mexico. These statistics are evidence that there are continuing contamination problems in the food import system and provide justification for the importance of the FDA not only to continue, but also to improve its sampling and testing protocols to detect adulterated food coming into the U.S. food supply.

If the U.S. food safety system is allowed to continue unchanged, there are likely to be major increases in the occurrence and size of food-borne outbreaks as U.S. food imports increase from countries in which risky food production, harvesting, and processing practices exist. This issue is among the most serious of food safety concerns confronting Americans for the foreseeable future. This book is the first to provide a comprehensive treatment of the microbiological food safety issues facing the United States from imported foods, and provides the justification for changes in the U.S. food safety net.

MICHAEL DOYLE
MARILYN ERICKSON

Imported Foods: Microbiological Issues and Challenges
Edited by Michael P. Doyle and Marilyn C. Erickson
© 2008 ASM Press, Washington, DC

Status and Projections for Foods Imported into the United States

W. J. Florkowski

INTRODUCTION

International trade in food expands the variety and lowers the cost of food, benefiting both the exporting and importing countries. Although the exact estimates of gains from food imports are lacking, the total welfare gains from imports equaled $260 billion in 2001 (Broda and Weinstein, 2005). The share of imported food has been climbing for the past two decades in the United States. The per capita food consumption grew from 1,800 lbs in the early 1980s to more than 2,000 lbs in 2000 (Jerardo, 2003), while the share of imported food in domestic consumption increased from 11% in 1990 to 15% in 2006 (A. Jerardo, personal communication). The fastest growth in consumption was reported, among others, for fruits and vegetables. The shifting composition of the American diet along with a higher demand for off-season fruit favors further growth in imported food (Jerardo, 2003). The top two food exporters to the United States are members of the North American Free Trade Agreement (NAFTA), Mexico and Canada (Table 1). In terms of the volume of imported food, five others among the top ten food exporters are countries located in Latin America, two in Asia, and Australia. Only one half of the top exporters in terms of volume is also among the top ten food exporters to the United States in terms of the export value. Table 1 shows also that the geographical direction of food imports to the United States varies with regard to the volume and value, which has potentially important implications in terms of food safety.

This chapter describes changes in imports of selected food categories, changes in the imported volume of specific foods, and reasons for the imports

W. J. FLORKOWSKI, Department of Agricultural and Applied Economics, University of Georgia, Griffin, GA 30223.

Table 1 Top ten food-exporting countries to the United States in volume and value in 2006[a]

Country	Rank by volume (metric tons)	Rank by value ($US)
Mexico	1	2
Canada	2	1
Costa Rica	3	–
Guatemala	4	–
Chile	5	6
China	6	9
Ecuador	7	–
Honduras	8	–
Thailand	9	–
Australia	10	3
Italy	–	4
France	–	5
The Netherlands	–	7
New Zealand	–	8
Brazil	–	10

[a]Source: Based on Foreign Agricultural Service data (USDA, 2007).

of red meats, seafood, fruits, vegetables, and nuts to the United States and identifies differences and similarities among exporting countries with implications for the risk of potential microbiological contamination in the domestic food supply. The description distinguishes among food categories and countries of origin with regard to the growth of food product imports. A description of the general conditions of the economic and social development in exporting countries is included as an indication of risk associated with the introduction of food-borne pathogens into the food supply system. Knowledge of countries of origin of food exports, the type of food being imported, the growth of the imported volume, its importance to the total supply on the U.S. market, primary ports of entry, and the seasonality of imports are relevant for an accurate assessment of contamination risk. Decision makers from public and private sectors gain valuable insights enabling the development of risk reduction strategies and enhancement of the integrity of the food supply system.

CONSUMPTION OF SELECTED IMPORTED FOOD GROUPS

Americans have been consuming an increasing share of foreign-grown foods. The average share of imports in total domestic consumption of food categories discussed in this section, i.e., red meat, seafood, fruits, vegetables, and tree nuts, steadily increased between 1980 and 2005. Among animal products, the

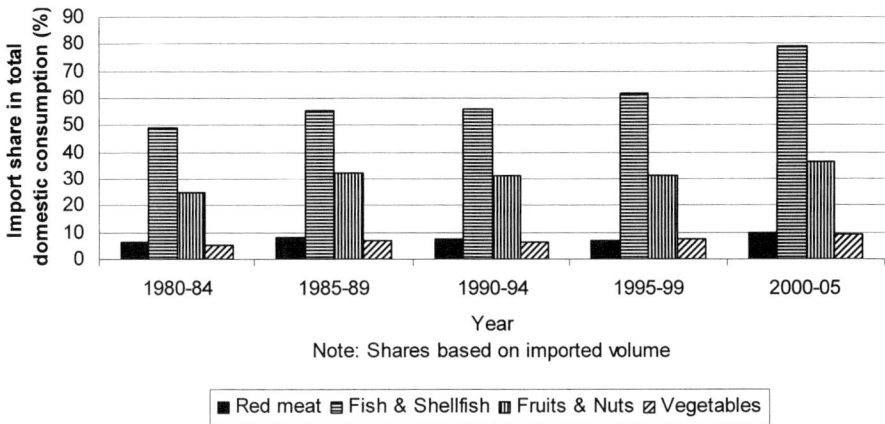

Figure 1 Average shares of selected imported foods in total domestic consumption between 1980 and 2005. Based on data from A. Jerardo (personal communication).

average share of red meat imports for the periods of 1980 to 1984 and 2000 to 2005 increased from 6.4% to 9.6%, respectively (Fig. 1). The share grew especially rapidly between 2000 and 2005 and is likely to increase more rapidly as the United States shifts a major portion of its use of corn from meat and poultry to biofuel production. The share of imports in the domestically consumed fish and shellfish category is much greater than in the case of red meat. In the period between 1980 and 1984 imports accounted, on average, for 49.2%, while in the period between 2000 and 2005 they accounted for 79.1%. The average share of imported fruits and nuts was 24% between 1980 and 1984, but it amounted to 36.3% of total domestic consumption between 2000 and 2005 (Fig. 1). The growth of imports in the domestic consumption of fruits held relatively stable during the 1990s but has rapidly increased since 2000. However, the most rapid relative growth in the share of imports in total domestic consumption is for vegetables. Imported vegetables represented an average of 5% in total domestic consumption from 1980 to 1984, but 9.3% between 2000 and 2005 (Fig. 1). The discussed food categories represent the high-value items. Moreover, many of them have been fully or partially processed; for example, fish and shellfish are imported as chilled filets or peeled shrimp.

The composition of imports and the share of individual foods has been changing over time. In the case of red meat imports, the average share of imported pork declined between 1980 and 2000, but the share of beef and, especially, lamb and mutton increased. Imported beef accounted for 14% of domestic consumption from 1996 to 2000 (Fig. 2), an increase of 40% since 1980 to 1984. The average share of imported lamb and mutton in total domestic consumption increased from 7.5% between 1980 and 1984 to 32.8%

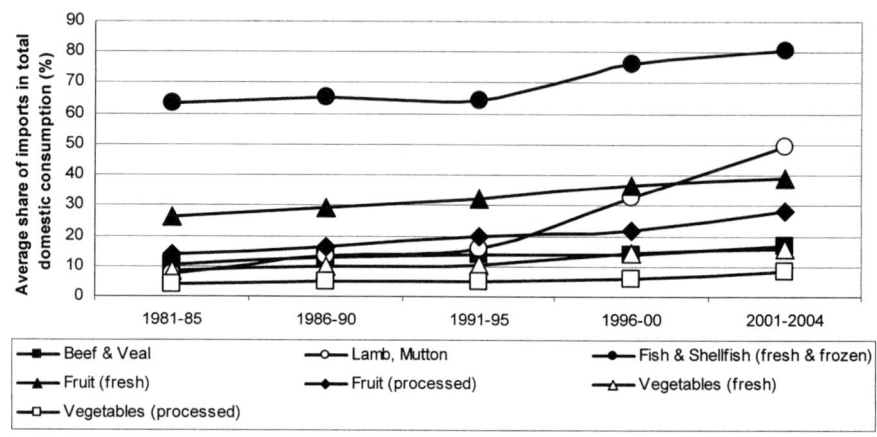

Figure 2 Average shares of selected imported food products in total domestic consumption (1981 to 2004). Based on data from A. Jerardo (personal communication).

between 1996 and 2000. Although the average share of imported fresh and frozen fish and shellfish as a percent of total domestic consumption increased from 63.3% from 1980 to 1984 to 76.3% between 1996 and 2000 (Fig. 2), the share of imported canned fish increased relatively faster from 25.6% to 36.2% between 1980 to 1984 and 1996 to 2000, respectively.

In the case of imported fruits, the average share of imported fresh fruit in total domestic consumption has been increasing more rapidly than in the case of processed fruit. The average share of imports between 1980 and 1984 was 26% of total domestic fresh fruit consumption, but 36.6% (40.7% share growth) from 1996 to 2000 (Fig. 2), while the similar shares for the same periods in the case of processed fruits were 13.8% and 30% (57% share growth), respectively. However, the fresh fruit and processed import share has been growing more rapidly since 2000, although the growth is concentrated on a limited number of fruits. The growth of the share of imported tree nuts in total domestic consumption has been also increasing, but the growth rate is rather small.

Imports of fresh vegetables represented on average 8.6% of total domestic consumption between 1980 and 1984 and 14.5% between 1996 and 2000, respectively (Fig. 2). The growth reflects the increased import of selected vegetables discussed later in this chapter. The growth in fresh vegetable imports contrasts with a relatively small growth in the case of processed vegetables, including dry beans and frozen fried potatoes. The average share of imported processed vegetables increased from 4.1% in the period between 1980 and 1984 to 6% in the period between 1996 and 2000 (Fig. 2).

Partially or fully processed imported food products provide some explanation of the observed increase in food imports share. Increased consumption of imported foods is associated with demand for convenience and changes in household incomes. However, other factors driving the demand for imported foods include preference for variety, produce availability extended beyond the harvest season determined by climatic conditions, competitive prices, and household size and demographic composition. The share of imported foods in total domestic consumption takes place despite the weakening dollar making every imported item more expensive to American consumers.

CHANGES IN IMPORT SHARES OF MEAT, FISH, SHELLFISH, FRUIT, VEGETABLES, AND TREE NUTS IN TOTAL DOMESTIC CONSUMPTION BETWEEN 2000 AND 2005

Food imports to the United States have accelerated since 2000, but the increase in the imported volume varies across specific products. For the majority of products considered in this chapter, the share showed a tendency to increase between 2000 and 2005, although the growth may have been variable. Increasingly, imported food is to some extent processed. This service included in the product simplifies handling by distributors and consumers because preliminary marketing functions, handling, and processing have been conducted at a location outside the United States, either in the growing area, at an export facility, or outsourced to yet another country prior to shipping to the United States. Meat, fish, and shellfish are most often subject to processing, while the handling of fresh fruits or vegetables may be limited to sorting, grading, and packing. Differences in labor costs between the importing and exporting country are the immediate reason to process food products outside the United States. An illustration of labor cost differences is provided later in this chapter.

Red Meat

The growth of red meat imports results from increased beef and lamb imports. The share of imported beef in total domestic consumption increased from 15.9% in 2000 to 19% in 2004, while the corresponding shares of imported lamb are 41.2% and 54.4%, respectively. Although the per capita lamb consumption is relatively small, beef is consumed in the largest volume among all meat kinds. The share of imported pork has declined, and the (likely) temporary decline is related to the trade dispute between the United States and Canada, the primary source of imported pork.

Fruits

Overall, the share of imported fresh fruit changed little between 2000 and 2005, from 38.2% to 39.7%, respectively. However, the changes have been

Table 2 Growth in imported volume of selected fresh fruits between 2000 and 2005[a]

Fruit	Change in imported volume 2000–2006 (%)	Share of imports in total domestic consumption in 2004 (%)	Primary exporting country	Per capita consumption of fresh fruit[b]
Citrus fruit				
Lemons	132[c]	7	Mexico	2.9
Limes	55[c]	98	Mexico	2.2
Grapefruit	89	3	Bahamas	2.6
Noncitrus fruit				
Raspberries	292[c]	Negligible	Mexico	0.3
Sweet cherries	364	3	Chile	0.9
Avocados	145	64[d]	Chile	3.3
Papaya	89	89	Mexico	0.9
Blueberries	73[c]	34	Chile	0.4
Cranberries	84	?	Canada	0.1
Peaches and nectarines	70[c]	9	Chile	4.8
Pineapples	99	79	Costa Rica	3.9[e]
Plums	48[c]	21	Chile	1.1
Grapes	30[c]	55[d]	Chile	8.7
Watermelons	86	13	Mexico	13.8
Mangoes and guavas	25	100	Mexico	1.9
Honeydew	0[c]	30	Mexico/Guatemala	1.9
Cantaloupe	−15[c]	32	Guatemala	10.1
Apples[f]	−4	7[g]	Chile	17.1

[a]Source: Calculations based on USDA data. Primary exporting country from fresh produce imports into United States, Produce Marketing Association. Change in volume imported from 2000 to 2006 calculated based on data available at http://www.fas.usda.gov/ustrade.

[b]Farm weight in pounds per capita per year in 2005 as listed at http://www.ers.usda.gov. Accessed 20 April 2007. Consumption figures rounded to only one digit after the decimal.

[c]Change in imported volume between 2000 and 2005.

[d]USDA-ERS (2007).

[e]The per capita consumption of fresh pineapple in 2002. USDA-ERS (2003).

[f]Apple imports fluctuate; for example, between 2000 and 2004, the imported volume increased by 27%.

[g]Share in 2005 (A. Jerardo, personal communication).

more dramatic in the case of specific fruits. Imported citrus accounted for 10.9% of total domestic consumption in 2000, but for 14.9% in 2004. Citrus imports have widened consumer choices by enhancing the variety of fresh citrus, but some citrus imports have replaced domestic production. Lemon imports increased by 132% (Table 2) and the share of imports was 7% in 2004. Mexico is the primary lemon exporter to the United States. Mexico is also the exclusive source of imported limes, supplying 98% of total domestic consumption. The imports of this fruit increased by 55% between 2000 and

2005 (Table 2). The Bahamas are the primary grapefruit exporter to the United States. Grapefruit imports increased 89% between 2000 and 2006, while their share in domestic consumption was 3% in 2004 (Table 2).

Among noncitrus fruit, the share of imported avocados amounted to 64% in 2005, and the volume imported increased 145% between 2000 and 2006 (Table 2). The share of imported fresh grapes in total domestic consumption increased from 44% in 2000 to 55% in 2004. Imported grapes are mostly from Chile, where the harvest season complements California's harvest. Fresh grape consumption has propelled that fruit to the top ten fruits in terms of their per capita consumption in the United States. The share of imported apples in total domestic consumption increased from 5% in 2000 to 7% in 2005, but the volume imported widely fluctuates from year to year in response to domestic supply-and-demand conditions. For example, in 2006, the imported volume of apples decreased in comparison with imports in 2000. Although tradition- ally apple imports originated in Canada (Carew et al., 2006), in recent years new marketing arrangements following the development of new varieties led to apple imports from Chile and New Zealand (U.S. Department of Agriculture, Economic Research Service [USDA-ERS], 2005) as well. In the case of watermelons, the imported volume has been increasing. Traditionally, domestic producers targeted the Fourth of July celebration as the beginning of the watermelon eating season, but unpredictable weather delaying the ripening disrupted the supply and encouraged imports. In 2004, 13% of do- mestically consumed watermelon was imported, primarily from Mexico, while the imported volume changed by 86% between 2000 and 2006 (Table 2). A rapid increase in imports was observed in the case of sweet cherries from Chile: 364% since 2000 (Table 2), although imports represent a small portion of the domestic consumption (3% in 2004). It is expected that because Canada (primarily British Columbia) has greatly expanded sweet cherry pro- duction planting, new varieties better withstanding shipping and later harvest dates than in the United States will increase exports in years to come.

Peach and nectarine imports increased by 70% since 2000 (Table 2). Their share in total domestic consumption is relatively small, 9% in 2004. These fruits are imported mostly from Chile after the domestic harvest and available to consumers in the winter. Plums, of which imports also originate in Chile, increased their import volume by 48% in the period under consideration, but their share in total domestic consumption is more than twice as large (21%) than the share of peaches and nectarines. Blueberry imports increased by 73% and represented 34% of total domestic consumption (Table 2). Fresh culti- vated blueberries in the off-season originate primarily from Chile, while some imports reach the United States from Canada, especially late in the season. Canada is also the primary source of wild blueberries, which are imported as processed product. The rapid increase in cranberry imports (73%) will continue if domestic supplies will not satisfy demand.

Among tropical fruit, the imports of mangoes increased by only 25% (Table 2), with Mexico being the exclusive supplier. Mangoes, the most often consumed fruit in the world, will likely expand beyond consumption by ethnic groups of consumers and reach mainstream consumers. The rapid increase in pineapple imports equaled 99% between 2000 and 2006 and is consistent with the large share, 79%, of imports in total domestic consumption. The increase in pineapple imports followed the currency crisis in Southeast Asia that erupted in 1997 and complemented the limited domestic supplies originating from Hawaii. Papaya imports increased by 89% between 2000 and 2006 and represented 89% of domestic consumption in 2004 (Table 2).

Imports of processed fruits include primarily fruit juices and wine. In 2004 about one fourth of total domestic wine consumption originated overseas. Among fruit juices, the share of imported apple, orange, and pineapple juices increased markedly between 2000 and 2004. In 2004, imported pineapple juice represented 87.9% of total domestic consumption, while apple and orange juice imports accounted for 75.2% and 25.7% of total domestic consumption, respectively. The Philippines, Thailand, and Indonesia are the primary pineapple juice suppliers, contributing 92% of pineapple juice imports to the United States (USDA-ERS, 2003). Among the top exporters, the Philippines accounted for 51% of the imported pineapple juice volume in 2002. China is the largest apple juice concentrate supplier to the United States. In 2004, the volume of Chinese apple juice accounted for 56% of total apple juice imports to the United States (Huang and Gale, 2006). Chile and Argentina are the other two large exporters of apple juice to the United States. Brazil is the primary supplier of orange juice, and the volume imported fluctuates in response to the domestic orange juice supply.

Vegetables

The imported fresh vegetable share was 13.8% in 2000 and increased to 16.9% in 2004. The growth of imported volume varied across specific vegetables, and the growth of imported volume did not necessarily imply a large share of total domestic consumption. A particularly high growth in imported volume since 2000 occurred with leafy fresh vegetables. The imported volume of spinach increased by 314% (Table 3) and represented 3% of domestic consumption. Head lettuce imports increased by 303% in the same period but represented only 1% of total domestic consumption. Although the share of domestic consumption of both vegetables is low, the rapid increase in the volume imported indicates that the supply system is now in place and a price increase in the

Table 3 Growth in volume of imported of selected fresh vegetables between 2000 and 2005[a]

Vegetable(s)	Change in imported volume 2000–2005 (%)	Share of imports in total domestic consumption in 2004 (%)	Primary exporting country	Per capita consumption of fresh vegetables[b]
Spinach	314	3	Mexico	2.3
Head lettuce	303	1	Mexico	20.3
Mushrooms	62	9	Canada	2.6
Onions/sweet onions	51	11[c]	Mexico/Peru	21.2
Squash	48	38	Mexico	4.7
Asparagus	48	67[c]	Peru	1.2
Eggplant	39.5	40	Mexico	0.9
Bell peppers	60[d]	34[c]	Mexico	6.6
Tomatoes	36[d]	34[c]	Mexico	20.6
Artichokes	28	58	Spain	1.2
Broccoli	28	7	Mexico	5.6
Cucumbers	28[d]	49[c]	Mexico	6.5
Cabbage	27	3	Canada/Mexico	8.1
Potatoes	26	5	Canada	43.8
Radishes	17	25	Mexico	0.5
Carrots and turnips	49[c]	7	Canada	8.5
Snap beans	10	11	Mexico	1.8

[a]Source: Calculations based on USDA data.

[b]Farm weight in pounds per capita in 2005 as listed at http://www.ers.usda.gov and http://www.fas.usda.gov. Consumption figures rounded to only one digit after the decimal.

[c]Share in 2005 (A. Jerardo, personal communication).

[d]Change in volume imported between 2000 and 2006, in percent. Calculations based on data available at http://www.fas.usda.gov/ustrade.

United States can result in a supply response by exporters. Both spinach and head lettuce have a short growing period, enabling foreign growers to be fairly elastic in their response to changing demand in the United States.

Two other products with high growth rates in volume imported are mushrooms and onions. Although imported mushrooms represented only 9% of domestic consumption, their imports grew by 62% between 2000 and 2005 (Table 3). Onions' share in domestic consumption is 11% and their imported volume grew by 51% between 2000 and 2005. Imported onions comprised primarily sweet onions shipped from Peru and dry onions from Mexico.

Sweet onion imports will likely continue to increase because consumers tend to prefer this onion to the traditional (more pungent) onion varieties and may replace some of the traditional varieties with sweet ones. Because the sweet onion does not store as well as traditional varieties, imports are necessary to maintain the supply of this product throughout the year.

Imported asparagus accounted for 58.3% of total domestic consumption of fresh and processed asparagus in 2000. Fresh asparagus imports increased by 48% between 2000 and 2005 and accounted for 67% of domestic fresh asparagus consumption. Peru is the primary supplier of asparagus to the United States. Per capita consumption of asparagus has been increasing and its imports are likely to grow in response to demand from retailers.

Bell pepper imports reflect the preferences of ethnically diverse consumers and increasing knowledge about pepper use in cooking. The imported volume of this vegetable increased by 30% between 2000 and 2005 (Table 3). In 2000, the share of imported bell peppers was 22.2% and increased to 34% in 2005. Imported cucumbers have been enjoying a large share in total domestic consumption, and the share grew in recent years from 42.4% in 2000 to 49% in 2005 (Table 3). The imported volume grew by 28% from 2000 to 2006 and is likely to continue to grow. Some of the imported cucumbers are grown in hothouses in Canada and the Netherlands and represent an addition to varieties already imported and produced domestically. This added variety is an attempt to encourage greater cucumber purchase, and consumers seem to respond positively or the imports would phase out.

Among fresh vegetables available in the off-season are squash and eggplant. The volume imported of these two items grew by 48% and 40% between 2000 and 2004, respectively (Table 3). New varieties of these two vegetables and the production of baby squash and baby eggplant widen consumer choices and open new buyer segments, which likely will contribute to continuing imports. Fresh tomato imports increased by 36% from 2000 to 2006 and represent slightly more than a third of domestic consumption (Table 3). The imports supplement domestic supply during winter, but the imports grew despite some restrictions protecting domestic producers. Imported tomato share in the domestic consumption was 34% in 2005 (Table 3). Tomatoes are among the most popular vegetables, and linking tomato consumption to health benefits is likely to contribute to increased future purchases.

Artichoke imports increased by 28% and the imported volume represents 58% of domestic consumption. However, in per capita terms, artichoke consumption is relatively low. Fresh broccoli consumption increased by 28% from 2000 to 2004, and imports represent only 7% of domestic fresh broccoli consumption. Radishes, however, accounted for 25% of total domestic consumption in 2005, while the volume imported grew by 17% between 2000 and 2004 (Table 3).

Imports of processed vegetables accounted for 6.2% of total imported vegetable volume in 2000 and 9.3% in 2004. The growth was generally uneven across processed vegetables imported in the largest quantities, including dry beans, tomatoes, and mushrooms. However, in the case of frozen fried potatoes the import share increased from 14.1% in 2000 to 19.7% in 2004. Among processed vegetables with an increasing share are frozen cauliflower and broccoli.

Imports of vegetable juices have been fairly steady and hovered around 8 million liters between 2003 and 2006. However, the imported volume tends to increase in the case of unfavorable domestic crop of vegetables used in juice production. Among major suppliers of vegetable juice, the imported volume from Mexico declined, although this country traditionally supplies additional quantities in the case of domestic crop failure. The European Union countries are also large suppliers of a variable quantity of vegetable juices. In recent years, the most rapid growth of vegetable juice imports has been recorded in the case of China, which did not export in 2002, but exported 1.6 million liters in 2006 (a 133% growth in volume over the quantity imported in 2005); Thailand, which increased its exports of vegetable juices by 717% between 2002 and 2006; and Ukraine, which increased its exports by 58% from 2002 to 2006. In general, the volume of imported vegetable juice shows little growth in recent years, but the origin of imported vegetable juice shifted from Mexico, Canada, and the European Union countries to China and Thailand.

Tree Nuts and Related Products

The share of imported tree nuts increased between 2000 and 2004 from 42% to 44%. Cashews are imported in the largest volume. Between 2000 and 2005 the imported volume of cashews increased by 41.8% after the 58.5% growth in the volume imported in the preceding 5-year period (Table 4). All domestically consumed cashews are imported. Traditionally familiar to Asian immigrants, cashews have become a popular nut, tasty and relatively inexpensive. The increasing imports of cashews coincided with the increased world production and the entry of Vietnam into the world cashew market. India has long been the primary supplier of cashews and, while the imports from Brazil were fairly stable, imports from Indonesia and Thailand have been rapidly increasing. Cashews require special handling that is labor intensive, and countries of Southeast Asia have a competitive advantage in this area.

Pecan imports originate almost exclusively in Mexico. Imported pecans are mostly shelled halves and pieces. Pecan imports from Mexico accelerated in the late 1980s and grew by about 13% between 1995 and 2000 (Table 4), but by another 37.4% between 2000 and 2005. Because of labor cost differences, many United States-based shellers ship their in-shell pecans for custom shelling in the maquiladora enterprises (temporary importation of American-grown in-shell pecans for the purpose of shelling, sorting, grading, and packing of shelled pieces and halves and shipping back to the United States) located across the border, and some in the pecan industry question the true volume of

Table 4 Volume of U.S. imports of selected tree nuts from top exporting countries for 1995, 2000, and 2005[a]

Commodity/ country	1995[b]	2000[b]	2005[b]	Shared imports in total domestic consumption in 2004 (%)[c]	Increase 1995 / 2000 (%)[d]	Increase 2000 / 2005 (%)[d]
Brazil nuts						
Brazil	14,414	19,125	19,248		33	10.6
Bolivia	4,333	9,004	9,725		108	8
Peru	1,343	1,666	6,677		24	301
Chile	1,050	446	1,071		–	140
Others	232	386	1,071		66	178
Total	21,371	30,627	36,927	100	43	21
Cashew nuts						
India	60,029	98,036	116,015		63	18
Vietnam	375	22,412	81,558		5,876	264
Brazil	52,515	58,123	57,230		11	–
Indonesia	1,197	2,342	3,292		96	41
Thailand	0	101	3,165		–	3034
Others	5,267	8,209	7,096		56	–
Total	119,383	189,222	268,358	100	59	42
Chestnuts						
Italy	6,372	6,744	6,243		6	–
China	123	631	1,686		413	257
South Korea	566	1,722	1,614		204	–
Portugal	149	306	158		105	–
Spain	67	96	95		43	–
Others	240	291	79		21	–
Total	7,516	9,789	9,875	100[e]	30	1
Coconut meat						
Philippines	77,231	89,882	83,873		16	–
Thailand	6,596	25,998	32,208		294	24
Dominican Republic	35,150	37,094	28,207		6	–
Mexico	5,861	9,459	17,638		61	87
Canada	1,771	3,496	3,307		97	–
Others	5,833	9,933	4,947		70	–
Total	132,441	175,861	170,181	100[e]	33	–
Hazelnuts (filberts)						
Turkey	14,005	10,555	8,414		–	–
Italy	61	60	2,691		–	4,385
China	0	0	1,119		–	–
Others	1,332	1,271	1,338		–	5
Total	15,398	11,886	13,562	69	–	14

Table 4 *(continued)*

Commodity/ country	1995[b]	2000[b]	2005[b]	Shared imports in total domestic consumption in 2004 (%)[c]	Increase 1995 / 2000 (%)[d]	Increase 2000 / 2005 (%)[d]
Macadamia nuts						
Australia	2,499	6,487	4,935		159.6	–
South Africa	737	3,257	3,982		341.9	22
China	81	33	2,427		–	7,255
Guatemala	818	1,007	1,940		23.1	93
Kenya	552	32	1,516		–	4,638
Others	1,277	1,080	3,151		–	192
Total	5,965	12,896	17,950	57	116.2	39
Pecans						
Mexico	57,305	64,588	88,727		12.7	37
South Africa	17	0	1,547		–	–
Australia	1,040	245	700		–	186
Thailand	1	1	381		–	38,000
Others	48	898	532		1,770.8	–
Total	58,410	65,732	91,886	47	12.5	40

[a]Source: Calculations based on USDA data.

[b]Volumes in thousands of pounds.

[c]Share of imports in total domestic consumption as the percent of total food disappearance as listed at http://www.ers.usda.gov/Data/FoodConsumption/FoodAvailSpreadsheets.htm#nuts.

[d]Only an increase in the volume imported is indicated and rounded to the nearest full number. A decrease in imports was not marked.

[e]A small amount of domestically produced chestnuts is marketed. Hawaii supplies a small amount of domestically grown coconut.

imports. However, many Mexican growers export pecans to the United States, where prices are higher than in Mexico. Because the commercial pecan production is located mostly in northern Mexico, one can expect that pecan imports will continue despite the likely increase in domestic Mexican consumption of these nuts.

Among tree nuts with a rapid growth rate in their share or volume imported are macadamia nuts. Although exports from Australia, where the nut is an indigenous species, have been taking place for years and declined recently, new exporters are rapidly increasing their presence on the nut market in the United States. Among new macadamia-exporting countries are the Republic of South Africa, which increased its exports by about 22% between 2000 and 2005. However, the entry of China and Guatemala with rapidly increasing volume exported to the United States is likely only the beginning of a continued import growth (Table 4). Imports from Kenya may signal another geographical shift in production.

In contrast to changes in macadamia imports, hazelnut imports have experienced a relative decline. A smaller imported volume from Turkey was not offset by increased imports from Italy or China (Table 4). However, the emergence of China as an exporter of hazelnuts is noteworthy because hazelnuts have not been exported from this country to the United States in the past. The export of hazelnuts from China may signal this country's entrance on the world hazelnut market following the pattern of entry on the peanut, walnut, and macadamia markets. While it may be too early to predict increased hazelnut exports to the United States, it is not impossible that China will gradually replace Turkey as the primary hazelnut supplier, especially if the hazelnut consumption increase in Europe and the Middle East places additional demands on Turkish suppliers.

The volume of imported Brazil nuts increased by about 20.6% between 2000 and 2005 after increasing by 43.3% in the preceding five-year period, 1995 to 2000 (Table 4). The volume imported from the two largest suppliers, Brazil and Bolivia, showed little gain between 2000 and 2005. Peru and Chile increased their volume exported by 140.1% and 177.5%, respectively, between 2000 and 2005.

Coconut meat imports have stabilized in recent years (Table 4). The Philippines remains the largest exporter of coconut meat to the United States, but Thailand and Mexico recently have rapidly increased their exports (Table 4). Both Thailand and Mexico experienced severe currency crises in the 1990s (Mexico in 1994/1995 and Thailand in 1997), which negatively affected domestic markets and stimulated the need to increase exports. The import growth in the volume shipped from Thailand reached 294.1% between 1995 and 2000 and nearly 24% between 2000 and 2005. The growth rates of imported volume from Mexico were 61.4% and 86.5%, respectively.

Chestnut imports increased by <1% between 2000 and 2005. Although the imported volume from Italy remained fairly flat between 1995 and 2005, China and the Republic of Korea rapidly increased their exports from 1995 to 2000 (Table 4). Imports from China increased by 413%, while imports from Korea increased by 204.2% during that period. However, as the volume imported from Korea declined slightly between 2000 and 2005, the volume imported from China registered 257.2% growth during the same period. Again, the emergence of China as the supplier of chestnuts signals a new trend and an increased role of this country as the chestnut supplier to the United States. Chestnuts imported from China are harvested from an indigenous chestnut variety and will likely continue. Unlike the American chestnut, the Chinese chestnut tree is resistant to the Dutch elm disease.

Tree nuts have received very favorable publicity in recent years with regard to their potential benefits in maintaining health or preventing disease, which

Table 5 Leading seafood exporters to the United States in 2004[a]

Country	Quantity imported, in tons	Value (in million $)
China	372,593	1,240
Canada	328,601	2,114
Thailand	295,341	1,355
Chile	143,917	668
Ecuador	103,364	455
Indonesia	97,154	634
Vietnam	81,950	557
Philippines	69,626	197
India	56,762	406
Taiwan	54,686	140
Mexico	50,357	433
Iceland	31,250	155
New Zealand	30,219	134
Argentina	28,827	78
Brazil	25,906	167
Russia	23,561	221
Venezuela	22,803	130
Norway	19,215	106
Bangladesh	18,531	178
Trinidad and Tobago	18,528	43
Total imports	**2,131,697**	

[a]Source: Based on Seafood Market Analyst—2004 Review available at http://www.seafoodreport.com.

has led to increased demand for nuts. Although tree nuts are a very heterogenous category in terms of taste, uses, and origin, their total consumption can be expected to increase in the foreseeable future.

Fish and Shellfish

There are many products falling in this category, and the fish and shellfish trade involves virtually all countries in the world. In this category the share of imports in the United States consumption is exceptionally high. Already in 2000 the share of imports in total domestic consumption was 68.3%. It has increased since then and reached 85.9% in 2005. Despite the increasing imports, an average American still eats less than the recommended quantity of fish (and a fraction of the consumed volume of red meat) and the growth of imports will likely continue.

Table 5 shows the leading exporters of seafood to the United States in 2004. The rankings change little from year to year. Any changes in ranking take place gradually over time. The most noticeable has been the advancement of China to the top exporter in terms of the volume of the United States imports. Other large exporters include Canada, Thailand, Chile, Ecuador, Indonesia, Vietnam, and the Philippines. In several of these countries, fish farming is as important or, in the case of exports, more important than wild catch. Thailand, Indonesia, China, Vietnam, and the Philippines are important exporters of various shrimp, some of which are commonly farmed. Chile and Norway are important exporters of farmed salmon.

The value of imported seafood varies. Some countries export to the United States premium fish and shellfish that sell at a high price on a per unit basis. For example, the volume imported from Canada was about 12% lower than the imports from China, but the value was almost 71% higher than the value of Chinese exports in 2004 (Table 5). Similarly, a relatively high-value seafood was imported from Indonesia, India, Mexico, Russia, and Bangladesh. High-value products can more easily absorb an additional cost of handling to enhance food safety than low-priced imports because the relative price increase is likely to be small.

The volume of imported seafood has been increasing, and domestic consumption is very dependent on imports. Among the top ten consumed fish and shellfish, only catfish is primarily supplied by domestic producers. Catfish consumption increased by 7% between 1992 and 2006 (excluding imported basa, tra, and other similar species, which the United States law prohibits being called "catfish"), placing it sixth among the top ten consumed fish and seafood products (Table 6). Catfish producers experienced pressure from imported Asian catfish and tamed the rising imports as a result of a dumping investigation. However, the exporters of various species of Asian catfish continue to supply frozen product at an unprecedented pace to the United States. In January 2007, four million pounds of frozen channel catfish were imported, representing a growth of 1,000% in comparison with January of 2006 (*The Catfish Journal*, 2007). Although the growth rate in subsequent months can decline, prior to that period the volume of imported frozen catfish from China grew at one million pounds a month and imports from other Asian countries, primarily Vietnam, also grew. Some American catfish producers expressed the view that they may be forced to supply exclusively fresh catfish only because of the competition on the frozen catfish market from Asia. At the International Boston Seafood Show in March 2007, among 54 companies offering processed catfish products, only four were from the United States.

Tilapia, the only other freshwater fish among the top ten types of fish and seafood eaten in the United States, is largely imported. Virtually unknown to American consumers a few years ago, it now places fifth among the top ten seafood products (Table 6). In 2006, the imported volume of tilapia reached

Table 6 Growth in weight consumed per person of selected fish and shellfish between 1992 and 2006 and the top ten seafood types by weight, per person, in 2006[a]

Fish/shellfish	Change in wt consumed per person (%)	Wt consumed in 2006 (lbs)
Shrimp	64	4.40
Tuna[b]	−17	2.90
Salmon	133	2.03
Pollock	33	1.64
Tilapia	NA[c]	1.00
Catfish[b]	7	0.97
Crab	100	0.66
Cod	−53	0.51
Clams	−15	0.44
Scallops	15	0.31

[a]Source: NMFS (2007b).

[b]Tuna refers to canned tuna only, while catfish consumption excludes similarly imported fish because, according to the United States law, it cannot be called "catfish."

[c]NA, not applicable. No measurable per capita consumption in 1992.

158,253 t (Josupeit, 2006a) and was 17.3% higher than in 2005 and 391% higher than in 2000. The imported volume suggests a per capita consumption of about 1.2 kg (2.7 lbs) per capita in live weight equivalent. Central and South American countries supply the largest volume of imported fresh tilapia fillets (Table 7). Tilapia imported from Ecuador, Honduras, and Costa Rica represented 92% of the volume of imported fresh tilapia in 2006. However, frozen tilapia and tilapia fillets represent the vast majority of imported tilapia and 85% of the imported volume is from China. Moreover, the growth in imported volume is primarily due to the increased volume shipped from China. In 2006, frozen fillets became the largest tilapia product category among all imported tilapia products. Food service is the primary recipient of imported tilapia and will likely remain so in the foreseeable future, as the wholesale prices of chilled tilapia fillets have decreased (Josupeit, 2006a).

Among saltwater fish, tuna remains widely popular and is the most often consumed fish (Table 6). Tuna is consumed mostly in canned form, but the frequency of fresh and frozen tuna steaks is increasing, especially as the ways of consumption, for example, sashimi, move from the ethnic into the mainstream cuisine. The tuna industry has been under pressure to innovate and go beyond the traditional canned product. Recently, a number of new products have appeared on the market using new, consumer-friendly packaging. The mercury scare contributed to extinction of the tuna canning industry in the United States, and most of canned tuna is imported. However, tuna imports

Table 7 The imported volume of tilapia products to the United States in 2006, by country[a]

Tilapia product	Country	Volume, in tons	Change since 2005 (%)
Frozen fillets	China	63,300	43.5
	Indonesia	7,106	10.5
	Taiwan	3,137	1.8
	Thailand	198	−77.2
	Panama	212	14.6
	Others	428	−35.3
	Total	74,381	33.7
Frozen	China	40,544	31.3
	Taiwan	18,260	−24.3
	Others	1,968	30.2
	Total	60,772	7.5
Fresh fillets	Ecuador	10,870	2.5
	Costa Rica	2,677	−28.3
	Honduras	7,250	10.3
	Brazil	1,018	5.7
	Others	1,285	66.7
	Total	23,100	1.6

[a]Source: Josupeit (2006a).

encompass several products (Table 8). The imported volume of frozen tuna declined by 81% between 2000 and 2005. In 2005, tuna and skipjack were imported in the largest quantity from Thailand, the Philippines, and Ecuador (Table 9). Skipjack imports typically peak in December each year, and the primary port of entry is New York. Yellow fin tuna imports are considerably smaller (Table 10), and the largest suppliers include Vietnam, Brazil, Trinidad, and Costa Rica besides the Philippines. Imported yellow fin tuna enters the United States primarily through Miami, and the volume imported tends to be the highest in the second quarter of the year.

Cod, once a commonly consumed fish, is eighth among the top ten fish and seafood products. Both tuna and cod consumption decreased between 1992 and 2006, and the decrease was especially large in the case of cod, whose consumption decreased by 53% (Table 6). The volume of cod caught has declined in recent years because of reduced availability.

In contrast, the consumption of salmon has increased dramatically, propelling this fish to the third most often consumed seafood (Table 6). Its consumption increased by 133% between 1992 and 2006, while the consumption of pollock increased by 33% in the same period, placing it fourth among the top ten seafood products in terms of consumed weight. Whereas pollock is

Table 8 The United States import volume of tuna by country and product in 2005[a]

Product	Country	Volume, in 1,000 tons
Frozen tuna	Indonesia	3.7
	Philippines	1.6
	Canada	1.0
	Others	4.2
	Total	10.5
Tuna loins	Thailand	8.7
	Trinidad and Tobago	13.4
	Fiji	14.5
	Ecuador	6.5
	Others	3.6
	Total	46.7
Canned tuna	Thailand	77.4
	Philippines	43.8
	Ecuador	15.5
	Indonesia	18.0
	Others	14.3
	Total	169.0
Tuna in pouches	Thailand	19.7
	Ecuador	13.6
	Others	2.7
	Total	36.0
Fresh tuna	Ecuador	0.7
	Vietnam	3.3
	Brazil	1.8
	Mexico	2.1
	Panama	1.7
	Costa Rica	1.6
	Trinidad and Tobago	2.9
	Others	11.8
	Total	25.3

[a]Source: Josupeit (2006b).

mostly wild caught, salmon comes in increasing volume from fish farming. Tables 11 and 12 list the imported volume of Atlantic salmon, both whole and filleted, in 2005. Atlantic salmon fillets are imported primarily through the port of Miami, but there is not a typical seasonal pattern in imports. Most whole Atlantic salmon imports enter the United States through Seattle, and the bulk of the volume is imported in the first half of the year, while the largest volume of the Chinook salmon (also imported through the port of Seattle) is imported

Table 9 United States imports of tuna and skipjack by country of origin in 2005[a]

Import origin	Volume, in 1,000 tons	Market share (%)	Value ($100s)	Unit value ($/lb)
Thailand	69,175	47.1	145,249	0.95
Philippines	43,241	29.4	77,357	0.81
Ecuador	13,555	9.2	29,577	0.99
Indonesia	10,974	7.5	24,219	1.00
Vietnam	7,915	5.4	15,913	0.91
China	997	0.7	1,882	0.86
Mexico	240	0.2	668	1.26
Japan	40	0.0	92	1.03
Singapore	15	0.0	32	0.94
Other	683	0.5	2,011	1.34
Total	146,837	100.0	297,000	0.92

[a]Source: U.S. Import Edition, vol. 11, no. 4. Available at http://www.SeafoodReport.com.

Table 10 United States imports of yellowfin tuna, fresh or chilled, except fillet, liver, or roe, by country of origin in 2005[a]

Import origin	Volume, in 1,000 tons	Market share (%)	Value ($100s)	Unit value ($/lb)
Trinidad	2,635	15.4	20,861	3.59
Philippines	2,545	14.9	16,403	2.92
Vietnam	2,378	13.9	16,500	3.15
Brazil	1,462	8.6	10,834	3.36
Costa Rica	1,207	7.1	9,696	3.64
Mexico	1,000	5.9	4,571	2.07
Sri Lanka	390	2.3	3,368	3.92
Thailand	207	1.2	1,428	3.14
Malaysia	141	0.8	1,314	4.23
Other	5,100	29.9	31,607	2.81
Total	17,064	100.0	116,583	3.10

[a]Source: U.S. Import Edition, vol. 11, no. 4. Available at http://www.SeafoodReport.com.

in the second half of the year. The whole, i.e., little processed, salmon was imported mostly from Canada, while salmon fillets were imported primarily from Chile. Canada and Norway also supplied filleted Atlantic salmon, but in relatively small volumes. China's role in salmon processing has been increasing and is reflected in U.S. export and import statistics, suggesting increased exports to China, but also imports from China, although a portion of the exports remains in the country, suggesting some domestic consumption. China has developed a highly competitive fish-processing industry (Franz, 2006).

Table 11 United States imports of atlantic salmon, farmed, fresh or chilled, except fillet, liver, or roe, by country of origin in 2005[a]

Import origin	Volume (1,000 tons)	Market share (%)	Value ($100s)	Unit value ($/lb)
Canada	57,457	89.3	270,848	2.14
United Kingdom	4,168	6.5	22,924	2.49
Chile	1,598	2.5	6,554	1.86
Norway	270	0.4	1,668	2.81
Ireland	150	0.2	1,259	3.81
Iceland	144	0.2	589	1.85
New Zealand	10	0.0	43	1.97
Other	540	0.8	2,036	1.71
Total	64,337	100.0	305,921	2.16

[a]Source: U.S. Import Edition, vol. 11, no. 4. Available at http://www.SeafoodReport.com.

Table 12 United States imports of Atlantic salmon fillets, farmed, fresh, by country of origin in 2005[a]

Import origin	Volume (1,000 tons)	Market share (%)	Value ($100s)	Unit value ($/lb)
Chile	83,387	87.4	433,046	2.36
Canada	9,655	10.1	72,157	3.39
Norway	1,324	1.4	9,293	3.18
United Kingdom	704	0.7	5,649	3.64
Ireland	144	0.2	1,618	5.11
Iceland	212	0.2	1,210	2.59
Other	29	0.0	162	2.56
Total	95,455	100.0	523,136	2.49

[a]Source: U.S. Import Edition, vol. 11, no. 4. Available at http://www.SeafoodReport.com.

Shellfish is very popular with American consumers, with shrimp topping the list of the ten most often eaten seafood items (Table 6). Consumption increased by 64% between 1992 and 2006. Shrimp imports have been increasing, and imported shrimp suppliers have broadened consumer choice by supplying new varieties. The imported volume of shrimp increased 11.6% between 2005 and 2006 (Table 13) and reached 590,299 tons (Lopez, 2007). Breaded frozen shrimp imports have been rapidly increasing since 2000, and the volume imported grew by 10.6% between 2005 and 2006 and represents 8.3% of the total imported volume of shrimp to the United States. Peeled frozen shrimp imports increased by 9.9% in 2006 compared with the previous year (27.5% of the total imported volume in 2006), while the imported

Table 13 Major shrimp exporters to the United States in 2006 and the change in imported volume[a]

Country	Volume imported in 2006, in tons	Change in imported volume between 2005 and 2006 (%)
Thailand	193,764	20.4
China	68,150	50.8
Mexico	59,363	19.7
Indonesia	58,729	11.6
Vietnam	37,078	−13.7
India	27,277	−23.6
Malaysia	20,349	17.6
Bangladesh	19,442	22.6
Venezuela	9,856	−13.3
Honduras	9,311	−11.4
Guyana	7,786	−9.5
Canada	7,050	−8.6
Peru	5,295	18.6
Nicaragua	4,816	−1.4
Panama	4,662	−21.4
Others	57,371	
Total	**590,299**	11.6

[a]Source: Lopez (2007).

volume of headless shell-on shrimp increased by 32.4% during the same period (43.3% of the total imported shrimp volume in 2006). The volume of other frozen preparations increased at the fastest rate, 39.5%, while other preparations declined slightly by 0.8%. Thailand supplies 33% of the imported volume of shrimp, followed by China, 12%, Indonesia, 10%, Ecuador, 10%, and Vietnam, 6%, respectively. The concentration of imports has increased, and the top five countries supply 71% of the imported volume of shrimp. Over time, the share of prepared shrimp has increased. For example, breaded shrimp represents 58% of imported shrimp volume from China, and imports from China account for 81% of the total imported volume of breaded shrimp. Breaded, frozen shrimp is also the second most important item in shrimp imports from Indonesia and Vietnam. Frozen shrimp prepared in ways other than breaded product is the most important item in imports from Thailand, which supplied 60.2% of the total imported volume of prepared frozen, other than breaded, shrimp (Lopez, 2007).

Shrimp imports from China increased by 50.8% in 2006 compared with the volume imported from that country in 2005 (Table 13). Among the top five shrimp exporters to the United States, only the volume imported from Vietnam decreased (−13.7%). China's volume growth benefited from

Table 14 United States imports of scallops, live, fresh or chilled, by country of origin in 2005[a]

Import origin	Volume (1,000 tons)	Market share (%)	Value ($1,000s)	Unit value ($/lb)
Canada	1,806	64.9	29,433	7.39
Mexico	927	33.3	5,425	2.65
China	21	0.8	116	2.50
Japan	19	0.7	374	8.76
Chile	9	0.3	122	6.45
Other	—[b]	0.0	8	30.00
Total	2,782	100.0	35,480	5.78

[a]Source: U.S. Import Edition, vol. 11, no. 4. Available at http://www.SeafoodReport.com.
[b]—, <1,000 tons.

importing the breeding stock of shrimp from Ecuador before a disease affected Ecuador's production. Periodic disease outbreaks in shrimp production have been reported in other countries as well, and these outbreaks will likely influence the rankings and the volume imported by country as well as the type of imported shrimp in the future. It should be noted that, over time, the top five shrimp-exporting countries have increased their share in the total imported shrimp volume to the United States.

Crab consumption increased by 100% between 1992 and 2006, but crab, clams, and scallops place relatively low in the ranking (Table 6). Scallop consumption increased by 15%, while clam consumption declined by the same percentage between 1992 and 2006. Scallops are imported almost exclusively from Canada or Mexico (Table 14), while China also has become a supplier with about 1% market share and a competitively priced product. The primary port of entry for scallops is Boston, and their volume peaks in the last quarter of the year.

SUPPLY-SIDE DRIVERS OF FOOD IMPORTS

Supply-side drivers include production costs, technological progress, government policies, and use of chemical inputs. In agricultural production, labor, capital, and land are the primary inputs and their cost ultimately determines the profitability of an enterprise. In fruit and vegetable production, especially for fresh consumption, labor costs during harvest represent a major expense to a grower. Because in the United States labor costs have been relatively high, production of foods that are labor intensive has been shifting to overseas locations. High labor costs are, to some extent, offset by hiring immigrant labor to harvest many crops, especially fruits and vegetables. Technological progress embodied in new improved varieties allows expansion of production into areas that earlier were unsuitable or marginal for a particular crop. With changes in government policies that allow new varieties to be patented, the progress in breeding improved fruit, nut, and vegetable varieties has been remarkable although uneven. Government policies of countries exporting food to the United States include economic growth strategies based on developing sectors focused on exports because of the low domestic demand. Finally, the use of chemical inputs affects the supply of food and is related to environmental concerns such as fertilizer and pesticide runoff or the regulation of fishing or the acquisition of multiple antibiotic resistance among harmful food-borne microbes. Countries with varying regulations for pesticide use that fail to monitor their application may inadvertently create advantages for producers, while compromising the integrity of the food supply system.

The Cost of Labor

Labor costs tend to be high in fisheries and fruit and vegetable production. Consequently, the production of fruits and vegetables that require particularly high labor input has been gradually relocating to other countries, which have a competitive advantage in production costs. The high labor cost of fishing results from labor regulations, for example, increasing costs of fishing under the flag of European Union countries, and regulations protecting economic interests or environmental zones, forcing fishing vessels to travel long distances and spend extended periods of time at sea.

Labor costs influence production costs of domestic food. There is ample empirical evidence that the documented and undocumented immigrants, primarily from Mexico, are important for agricultural production across the United States (Lewis, 2007). It has been estimated that 42% of all farm workers in the United States are foreign and 52% of them are in the country illegally (Mehta et al., 2000). Many immigrants harvest fruits and vegetables, filling the gap that could not be readily filled with domestic labor because low-skilled Americans shun work in agriculture (Trejo, 1998). Among Mexican immigrants, two-thirds have not completed a high school education and many never attended a high school. A low level of education translates into a low level of skills leading to low wages. It is possible that without immigrant labor in agriculture, the harvest of many fruits and vegetables would have been disrupted and American farmers would have been unable to effectively compete. For example, foreign workers accounted for 75% of hired workers in Florida (Emerson and Roka, 2002).

Information about the wages paid to agricultural workers in the United States and in many countries exporting food to the United States is fragmentary. However, the difference in labor costs can be illustrated with an

Table 15 Indexes of hourly compensation costs for production workers in manufacturing in selected countries in 2000 and 2005[a]

Country	2000	2005
North America		
United States	100	100
Canada	84	101
Mexico	11	11
South America		
Brazil	18	17
Oceania		
Australia	73	105
New Zealand	43	63
Asia		
Japan	112	92
Korea, Republic of	42	57
Taiwan	32	27
Europe		
Italy	70	89
The Netherlands	98	135
Norway	115	166
Portugal	23	31
Spain	54	75
United Kingdom	86	109

[a]Source: U.S. Department of Labor (2006).

index calculated for the labor costs in manufacturing by using data available for several countries (Table 15). Assigning the index value of 100 to the labor costs in the United States, the comparable costs in Canada are slightly higher, but those in Mexico are a small fraction of the United States costs. Indeed, the labor costs in Mexico are about one third lower than in Brazil in the manufacturing sector. Index values for both of these countries likely reflect the general labor costs in other countries in Central and South America. It is plausible that labor costs in the fruit, vegetable, and tree nut sectors of South American countries would be similar in relationship to such costs in the United States, if not lower.

Among Asian countries, labor costs in China are relevant in food production. The large rural labor pool allows production costs to remain low (Huang and Gale, 2006). Although only fragmentary information about monthly and daily wages in selected areas and jobs is available, Huang and

Gale indicated $60 to $75 per month and less than $2 per day as wage rates in China's rural areas. It is likely that costs have increased since this observation was made, but labor costs remain very competitive in rural China. In the case of other Asian countries, the index labor costs in manufacturing is also substantially lower than in the United States with the exception of Japan, where the index value is 92 (Table 15). Among the countries of Oceania, Australia's labor costs are higher than those of the United States, but New Zealand reports labor costs considerably lower. Among the European countries, the index value is highly variable. Portugal reports the lowest value, 31 points, while Norway's labor costs in manufacturing are the highest at 166. Spain and Italy show a cost of labor below that in the United States.

This limited overview of labor costs, using the manufacturing sector as the example, indicates in which countries, with a comparative advantage in labor costs, the greatest manufacturing is likely to be located. Therefore, food exports to the United States, including fruit and vegetable shipments, are likely to continue from Central and South American countries and the majority of Asian countries. Food exports from Canada and high-cost European countries are driven by factors other than labor costs alone. Food quality, absolute advantage in its production, or other factors (e.g., cultural) are reasons for the observed imports to the United States and may be vulnerable to changes in relative prices, variations in incomes of American consumers, or publicity influencing food fads, attitudes toward products or exporting countries, and reports of food-associated disease outbreaks.

Technology Development

Technological progress further encourages relocation of production. New technology embodied in new varieties adapted to previously marginal growing conditions expands the supply potential. For example, the release of cultivated blueberry varieties with low-chilling requirements allow for the expansion of cultivation range to areas with considerably milder winters than earlier varieties (Carew et al., 2005). Cultivated blueberries are now produced in regions of Chile and South Africa, while tests are being conducted in Mexico, Spain, and several Asian countries. The expanding blueberry production has established a year-round supply system of fresh cultivated blueberries.

Increased volume of imported fresh vegetables is also driven by the transfer of technology to other countries, often supported by American capital, where production takes place during the off-season in the United States. Corporate decisions of major supermarket chains have influenced the fresh fruit and vegetable imports by promoting the year-round availability and encouraged deliveries of the demanded volume and quality outside the harvest season in the United States. The creation of the supply chain was possible through innovation in shipping and handling, the development of new packaging and equipment, and protocols of handling fresh fruit and vegetables.

Government Policies

A major force behind the surge of food exports to the United States has been export-oriented growth policies adopted by less developed countries. A cluster of the Southeast Asian countries, including Malaysia, Indonesia, Thailand, Vietnam, and the Philippines, has been pursuing an export-led growth strategy for an extended period of time. This strategy, which may have contributed to the financial and currency crisis in 1997, has become even more important after the crisis because of the contraction in domestic demand. Thailand, Indonesia, Vietnam, and the Philippines are important seafood exporters to the United States. The Philippines and Vietnam are large exporters of coconuts and cashews, respectively. Thailand pursues a policy of diversifying its fruit and vegetable production aiming at export markets. Recently, Thailand has been promoting tropical fruit at major trade shows in the United States, attempting to familiarize the industry with rambutan, mangosteen, and other tropical fruit.

China has become an increasingly important food exporter to the United States. Its agricultural sector lags behind the development of its manufacturing sector, which attracts a sustained flow of foreign direct investment. Consequently, its farm sector seeks increased returns by exporting seafood, fruits, vegetables, and nuts because importers offer higher prices than domestic buyers. China is highly opportunistic in its approach to trade. It exploited the 1997 to 1998 recession in Southeast Asia resulting from the currency and financial crisis by promptly shipping products whose supply was disrupted. Given the competitively priced labor and the export-led growth strategy (including the exchange rate policy), Chinese food exports are likely to continue to increase in the foreseeable future despite the growing domestic demand.

In Central America, exports of macadamia nuts, fruits, and vegetables have been increasing, largely stimulated by domestic policies, low labor costs, and assistance from international economic organizations. Honduras, Costa Rica, Guatemala, and other countries have greatly increased exports to the United States during the past decade. Because their economies depend heavily on agriculture, policies supporting food production for overseas markets are unlikely to change in the foreseeable future, and their proximity to the United States offers a large and readily accessible market.

Pesticide Use and Environmental Concerns

Raising fruits, vegetables, and nuts requires chemical inputs because of pest pressures in many regions. Changes in labeling requirements and testing and the banning of some of the ingredients in pesticides have added to production costs. The partial ban on organophosphates, which benefited the environment, erased the competitive edge fruit growers enjoyed in some areas. Although many countries ban or restrict the use of many pesticides, the enforcement may be lacking. Samples of produce selected for testing have led to occasional rejection of the shipment, but the testing involves only a small fraction of imported foods.

Shifting production overseas poses food safety challenges by lengthening the supply lines, adding multiple handling points, and transferring process control away from an established system to an organization that evolves as trade expands. An evolving system may be incomplete because all the necessary institutions servicing the system have not been established or have not reached the necessary level of expertise.

DEMAND-SIDE DRIVERS

The increased demand for imported food, both fresh and processed, results when such foods meet the expectations of convenience, taste, and quality at a competitive price. Changes in the amount of leisure time and disposable income have influenced the demand for services included in food products such as full or partial preparation, eliminating or reducing the need for meal preparation at home. Ethnic diversity and increased migration from non-European countries have led to the introduction of new foods into the mainstream American diet and have created niche markets for fruits, vegetables, tree nuts, and seafood. The shrinking size of the American household and the increasing share of single-person households has altered food consumption behavior. The improving level of educational attainment, broadened leisure choices, and richer travel experiences have also influenced changing demands for fresh and processed food.

Changes in Household Environment and Behavior

Increased leisure has not led to increased allocation of time to food shopping and cooking. Between 1965 and 2003, the average market work week decreased from about 52 h to 38 h for highly educated men (Aguiar and Hurst, 2006). During the same period, leisure time increased for men and women regardless of the educational attainment level by up to 8 h per week.

The average household size decreased from 3.35 in 1960 to 2.57 in 2005 (U.S. Department of Commerce [USDC], 2006). Single-person households accounted for 26.6% of all households in 2005, a doubling of the 13.1% share reported in 1960 as their number increased almost fourfold during the period under consideration. Although family size has decreased, the median income per family increased from \$30,374 to \$54,061 in real dollars (in 2004 dollars) between 1960 and 2004 (USDC, 2005). Moreover, the share of disposable personal income spent on food purchase declined from 17.5% in 1960 to 9.9% in 2005 (USDA-ERS, 2006a). An increasingly larger portion of food expenditures is allocated to the purchase of food away from home. Between 1960 and 2005, this share almost doubled from 26.3% to 48.5% (USDA-ERS, 2006b). Spending on food sales away from home grew slightly more rapidly than the family real income.

Increasing the opportunity cost of time (i.e., the cost of 1 h of leisure increases as wage rates increase) and the relative decrease of food prices compared with other goods and services discourage home meal preparation. The need to efficiently use time promotes demand for convenience. Two trends illustrate this phenomenon. Shellfish preparation requires skills, while eating shellfish away from home offers a good culinary experience without the trouble of cleanup. Another example of demand for convenience is fruit selection. Apples, bananas, and citrus are among the most commonly consumed fruits because they are easy to eat, accessible, and competitively priced. Fruits that gained share in consumption in recent years, i.e., grapes, strawberries, and blueberries, meet the expectations of convenience in consumption.

High-income households differ in their consumption pattern from low-income households. The association between income and food consumption is pointedly illustrated by the Engle function. The Engle function describes the consumption of food and various food categories in response to changing real income. The consumption of some food categories increases (e.g., dairy products, meat, and seafood) whereas consumption of other categories decreases (e.g., starch sources such as bread or potatoes).

Educated consumers make different food choices, including eating more seafood, fruits, and vegetables. An increase in the average educational attainment level stimulates the demand for variety. Tourism is especially conducive to familiarizing the public with new fruits, vegetables, nuts, and seafood. Once unknown or ignored dishes, for example, sushi, sashimi, salsa, chutney, or tabbouleh, prepared using fresh seafood, vegetables, herbs, or fruit, have become increasingly common menu items.

Ethnic communities create niche markets for imported seafood, nuts, fruits, and vegetables. Hispanic and Asian Americans encourage the importation of food items consumed in their home countries. Because these minority groups are growing at rapid rates and are increasingly heterogeneous, it is likely that new niche markets for seafood, fruits, vegetables, nuts, and herbs will continue to emerge. Studies suggest that a particularly large unsatisfied demand is in the area of fresh fruits and vegetables (Batres-Marquez et al., 2003).

GENERAL ECONOMIC AND SOCIAL CONDITIONS IN MAJOR FOOD-EXPORTING COUNTRIES

The enforcement of domestic food safety regulations, international food security agreements, or the execution of contract specifications is linked to the level of development of domestic institutions enforcing regulations and monitoring the food supply system. This section presents several general measures developed by international organizations to address selected aspects of economic and social development. It is commonly accepted that countries with high levels of socioeconomic development are reliable trading partners and credible enforcers of the law and regulations. However, this general tendency is still subject to variations across countries that may fit into one category of measurement but lag behind in another measure.

Per Capita Income

Per capita income is the common measure implying the lack of development. This measure can be used to rank countries in terms of their ability to monitor and enforce food safety regulations agreed upon in multi- or bilateral international agreements or specified in the export–import contracts. A suitable per capita income measure is the Purchasing Power Parity (PPP) income because it accounts for variation in domestic price levels resulting from differences in consumed goods and services, natural resource endowment, and other factors. The general level of prices in an economy is linked to wage rates and the two tend to move in the same direction. Table 16 shows the PPP income for selected major exporters of produce, seafood, and nuts to the United States. With the exception of Canada, Japan, Italy, Spain, the Republic of Korea, and Chile (ranked 58), other major exporters of seafood, fruits, vegetables, or nuts to the United States do not rank among the top 75 countries in the world.

The PPP ranking of food exporters to the United States indicates possible economic reasons for the trade. First, the PPP ranking accounts for the costs of living, suggesting that a country ranked relatively low is characterized by low costs of living. Low living costs translate into low wages and low wages help the competitive advantage in the production of foods where labor inputs are a substantial portion of total production costs. Second, countries with a relatively high PPP income export foods to the United States because of a comparative or absolute advantage. Third, some countries export seafood to the United States because they fish closer to American harbors than harbors of their own or other countries and the transportation costs and the domestic demand would not ensure comparable profits. In most cases, this pattern of fishing and selling seafood is also associated with labor costs and the limited storage space on a fishing vessel.

Table 16 Selected development measures for major food exporters to the United States[a]

Country	Purchasing Parity Power per capita income (PPP) ($)		Transparency Index		Economic Freedom Index	
	2006 rank	PPP income estimated	2005 rank	Index	2006 rank	Index
Central America						
Costa Rica	62	12,000	51	4.2	46	2.69
Guatemala	106	4,900	117	2.5	74	3.01
Honduras	130	3,000	103	2.6	102	3.28
Nicaragua	129	3,000	107	2.6	80	3.05
South America						
Argentina	51	15,000	97	2.8	107	3.30
Bolivia	128	3,000	117	2.5	67	2.96
Brazil	74	8,600	62	3.7	81	3.08
Chile	58	12,700	21	7.3	14	1.88
Colombia	177	8,400	55	4.0	91	3.16
Ecuador	112	4,500	117	2.5	107	3.30
Peru	94	6,400	65	3.5	63	2.86
Venezuela	93	6,900	130	2.3	152	4.16
Africa						
Kenya	172	1,200	144	2.1	94	3.20
Mozambique	166	1,500	97	2.8	113	3.35
Republic of South Africa	57	13,000	46	4.5	50	2.74
North America						
Canada	11	35,200	14	8.4	12	1.85
Mexico	66	10,600	65	3.5	60	2.83
Europe						
Iceland	8	38,100	1	9.7	5	1.74
Italy	25	29,700	40	5.0	42	2.50
Norway	4	47,800	8	8.9	30	2.29
Russia	61	12,100	126	2.4	122	3.50
Spain	28	27,000	23	7.0	33	2.33
Turkey	71	8,900	65	3.5	85	3.11
Asia						
Bangladesh	144	2,200	158	1.7	141	3.88
China	85	7,600	78	3.2	111	3.34
India	125	3,700	88	2.9	121	3.49
Indonesia	124	3,800	137	2.2	134	3.71
Japan	14	33,100	21	7.3	27	2.26

(continued)

Table 16 Selected development measures for major food exporters to the United States[a] *(continued)*

	Ranks					
	Purchasing Parity Power per capita income (PPP) ($)		Transparency Index		Economic Freedom Index	
Country	2006 rank	PPP income estimated	2005 rank	Index	2006 rank	Index
Malaysia	60	12,700	39	5.1	68	2.98
Philippines	137	4,700	117	2.5	98	3.23
South Korea	34	24,200	40	5.0	45	2.63
Taiwan	27	29,000	32	5.9	37	2.38
Thailand	66	9,100	59	3.8	71	2.99
Vietnam	127	3,100	107	2.6	142	3.89
Australasia						
New Zealand	30	26,000	2	9.6	9	1.84

[a]Source: Transparency Index (http://www.transparency.org/cpi/2005.10,18.cpi.en.html). The Economic Freedom Index as published by the Wall Street Journal and the Heritage Foundation (http://en.wikipedia.org/wiki/Index_of_Economic_Freedom). The PPP income per capita as updated in March 2007 on the CIA World Factbook (http://www.cia.gov/library/publications/the-world-factbook/index.html).

The Transparency and the Economic Freedom Indexes

The categorization of countries exporting food to the United States serves the purpose of safeguarding the food supply and enhancing the trust of consumers in the quality of available foods. Because the enforcement of food safety regulations is increasingly important for the integrity of the overall food supply system, the development of a domestic regulatory system in the exporting country is essential. Although adoption of regulations is monitored, the essential step of any regulation is its implementation and execution. The Transparency Index (TI) and the Index of Economic Freedom (EFI) indirectly account for the systems of laws and regulations guiding food exports and are indicators of high-risk food exporters. Knowledge of the type of imported food and the country of origin can be used to anticipate the risk of importing contaminated food and guide inspection and testing of the incoming shipments.

Transparency International is an organization that has been calculating the Transparency Index. The rankings for 2004 were based on a public opinion survey of >50,000 individuals in 63 countries (Transparency International, 2004) and define corruption as "the abuse of entrusted power for private gain." Transparency within an economy determines the access and quality of information available to sellers and buyers. Lack of transparency reduces market information and creates systemic risks leading to a potentially inefficient market discovery process, thereby increasing costs of doing business. Corruption further distorts market signals. Among food exporters to the United States that

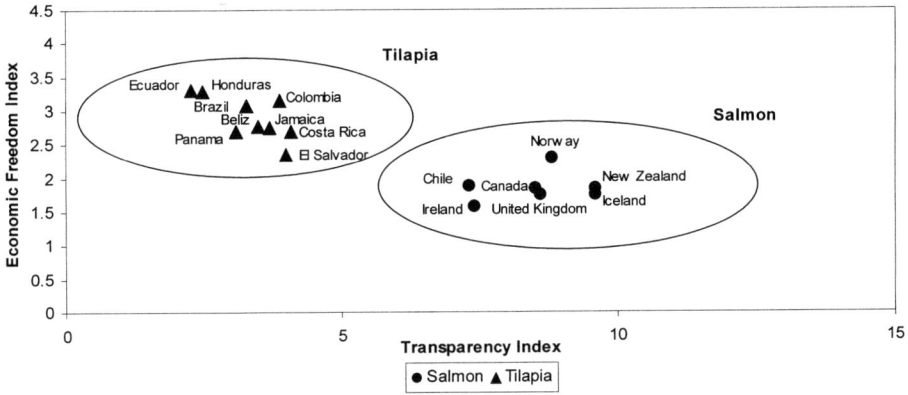

Figure 3 Transparency and economic freedom in countries exporting Atlantic and Chinook salmon and tilapia to the United States in 2006. Source: Transparency Index values obtained from Tranparency International's 2004 Annual Report (http://www.transparency.org). Economic Freedom Index values are as published by *The Wall Street Journal* and the Heritage Foundation (http://en.wikipedia.org/wiki/Index_of_Economic_Freedom).

consistently have been associated with a high level of corruption are El Salvador, Guatemala, and Honduras (Jansen, 2006).

The EFI is an indicator of the ease with which an entrepreneur can operate in a country (EFI, 2007). Specifically, the EFI considers the extent to which "individuals are free to work, produce, consume, and invest in any way they please, while protected by the state and unconstrained by the state" (EFI, 2008). The EFI may be linked to the Transparency Index because, in the process of administratively channeling economic activity, a government creates a risk of increased lack of transparency. In the case of food exports, the lack of transparency and the absence of economic freedom increase the risk of lapsed regulatory oversight. Consequently, regulations pertaining to food safety, including quality certification, may not be enforced on a consistent basis and result in increased risk to consumers.

Figure 3 illustrates the relationship between the TI and the EFI by plotting rankings of both indexes for countries exporting salmon and tilapia fillets. The exporting countries form two clearly distinguishable clusters showing that tilapia exports originate in countries with less transparency and economic freedom than salmon exports. The two groups of countries vary also in terms of their PPP income. Flawed inspection, faulty interpretation of food-handling rules during shipping, or inattention to food safety rules may increase the risk of incidental bacterial food contamination and can be tied to rankings accounting for transparency and economic freedom. Economic incentives to ensure export market access and the pressure to execute an order

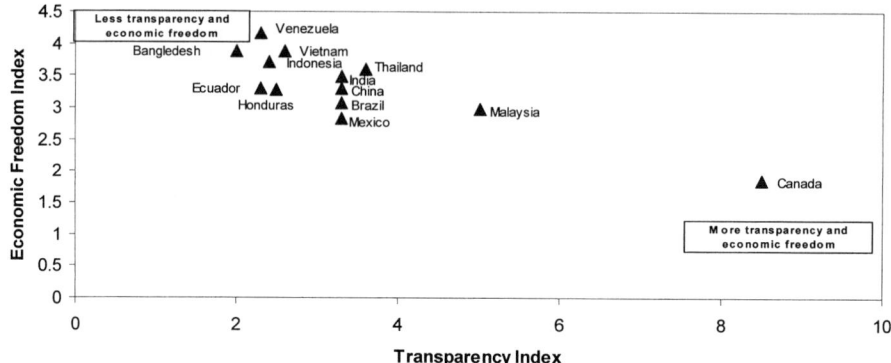

Figure 4 Transparency and economic freedom of major farmed and wild-shrimp exporters to the United States, 2006. Source: Transparency Index values obtained from Tranparency International's 2004 Annual Report (http://www.transparency.org). Economic Freedom Index values are as published by *The Wall Street Journal* and the Heritage Foundation (http://en.wikipedia.org/wiki/Index_of_Economic_Freedom).

in a timely fashion or eliminate potentially costly packing or handling procedures (e.g., temperature regime violation) create preconditions for payoffs to inspectors and other officials in charge of supervising and monitoring the process. Figure 4 illustrates the distribution of major shrimp-exporting countries according to their scores in transparency and economic freedom. Because a variety of farm-raised and wild-caught shrimp are imported, the countries on the plot are stretched from the upper left corner (implying a less transparent and free economy) to the bottom right corner (suggesting a more transparent and free economy). The exporting countries cluster in the upper left corner of Fig. 4, indicating rather untransparent and restricted economies. Such economies may have an insufficiently developed system for enforcing food safety regulations, leading to inconsistent monitoring and omission of incidents of food contamination.

Figure 5 provides another illustration of the association between transparency and economic freedom indexes. Two clusters of countries represent exporters of cashew nuts and chestnuts. Only China, among chestnut exporters, is closer to positions occupied by cashew exporters than to chestnut exporters. But China is a newcomer on the U.S. chestnut market, while Italy has long traditions in exporting chestnuts. As in the case of salmon and tilapia imports, among nut exporters, countries can be classified between those with less transparency and economic freedom and exporters where transparency is high and economic activity free of unnecessary regulations.

The consideration of transparency, economic freedom, and other measures of the country's overall development level is relevant because the overall

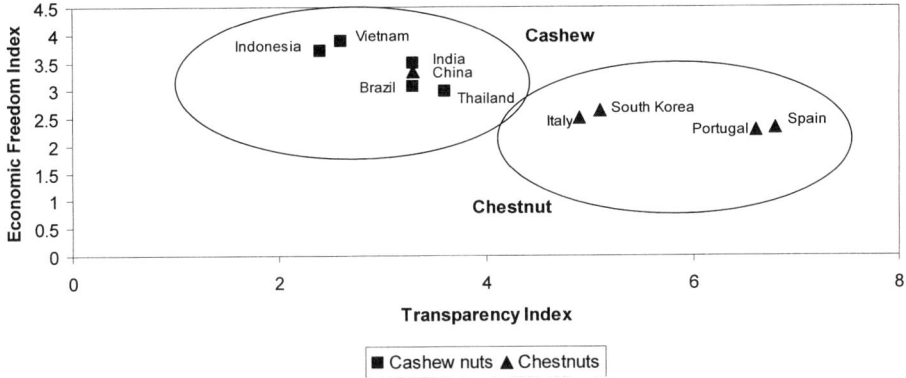

Figure 5 Transparency and economic freedom in countries exporting cashews and chestnuts to the United States, 2006. Source: Transparency Index values obtained from Tranparency International's 2004 Annual Report (http://www.transparency.org). Economic Freedom Index values are as published by *The Wall Street Journal* and the Heritage Foundation (http://en.wikipedia.org/wiki/Index_of_Economic_Freedom).

development involves the establishment and performance of institutions facilitating trade. Underdeveloped or missing institutions create an institutional vacuum, which allows the exploitation of the situation by entrepreneurs who disregard requirements or procedures, increasing the cost or causing delays in trade. Avoidance of regulatory compliance leads to attempts to bribe officials, which in countries with relatively low PPP income (and low labor costs) is a potentially attractive way to supplement a salary. In a number of countries, including food exporters to the United States, enforcement institutions lack skilled professionals (Messick and Kleinfeld, 2001). The eventual result of illegal behavior could be compromised food safety, which becomes evident in the importing country after the consumption of contaminated food.

PROJECTIONS

The short-term projections regarding vegetable imports were quite accurate in the past, but fruit import predictions were less accurate, especially in the category of fresh and frozen fruit (USDA, 2007). In the short term, the volume of imported produce will be affected by the changes in the dollar exchange rate, which is expected to increase the cost of imported food in the immediate future.

In the short term, the growth of the imported volume of horticultural products will continue at the recently observed pace. At the time of writing this chapter, USDA forecasted that in 2007 the volume of imported fresh fruit would increase by 6.3%, processed fruit imports would remain unchanged compared with 2006, fresh vegetables would increase by ≈7%, and processed vegetable imports would increase by ≈8% (Lin, 2004). Horticultural imports from Mexico and Canada, both NAFTA members, would increase by ≈7% in 2007. Imports would also increase from Central and South American countries and the Caribbean.

The long-term projections suggest a relatively more rapid increase in the consumption of seafood, fruits, vegetables, and possibly nuts and seeds than red meat because of the combined effects of increasing incomes, aging population, and the growth of ethnic groups (Ballenger and Blaylock, 2003). Lin et al. (2003) projected the long-term increase in consumption of vegetables of 20% between 2000 and 2020, a citrus fruit consumption increase of 27%, and a comparable growth in the consumption of apples. In the vegetable category, the growth of lettuce and "other vegetables" was anticipated to be more rapid than that of other types of vegetables. The predictions in total and per capita consumption emphasized the interaction of increasing income and eating out, which led to a decrease in fruit consumption, and knowledge. Knowledge seems to be a key factor influencing food choices, especially fruits and vegetables, yet knowledge is a complex factor shaped by formal education, experience, traditions, habits, etc. Overall, the predicted changes in consumption of fruits and vegetables suggest that a large portion of the consumed volume will be imported because the domestic supply will be limited by production factors and climate.

Tree nut consumption is likely to continue to increase, stimulated by research linking nut consumption to benefits in disease prevention and health maintenance. The per capita nut consumption has been increasing and is based mostly on domestic nut production, but imported nuts add variety to available choices and continue to be eaten by ethnic groups. Consumer studies indicate that consumers either eat many varieties of nuts or seldom purchase nuts at all (Park and Florkowski, 1999). Given the taste and price, cashews and Brazil nuts will likely increase in imported volume, while macadamia imports will likely increase if production in Central American countries increases.

Seafood imports will continue to grow because the demand for fish and shellfish is unlikely to decline. Between 1985 and 1997, seafood consumption in the United States increased at 1.5% annually (Msangi and Rosegrant, 2005). During the same period, the per capita consumption actually decreased by 0.5 lb. Since then, however, the per capita consumption increased by 1.3 lbs, reaching 16.2 lbs, slightly less than the record 16.6. lbs in 2004 (National Marine Fisheries Service [NMFS], 2007a). A 2% increase in the consumed volume of seafood in the United States equals about 5,200 tons

(11.5 million pounds) to 6,500 tons of imported seafood, assuming an 80% share of imports in total domestic consumption. Lin et al. (2003) project the increase in seafood consumption to increase by 30% in away-from-home consumption and by 23% in at-home consumption between 2000 and 2020. By using their projections, the expected per capita seafood consumption could reach ≈20 lbs in 2020 and will require a substantial increase in the imported volume of seafood.

The imported seafood will likely originate from aquaculture rather than wild catch. Increased imported volume from aquaculture will lead to the concentration of imports on a few species and small quantities of a variety of seafood types. Therefore, the management of risk of contaminated seafood will need to account for this pattern of imports. Shrimp production largely originates in the Pacific and Indian Oceans and has been growing at a more rapid rate in the Indian than in the Pacific Ocean. Preliminary figures for the imported volume of shrimp in January 2007 already indicate a substantial increase in imports from Thailand and China. Aquaculture shrimp production carries the risk of disease outbreaks that in the past affected Thailand (1996 and 1997), China (1993), and Ecuador (1999). Such outbreaks are likely to occur again and, given the importance of shrimp in seafood consumption in the United States, food service may be looking for alternative suppliers or alternative products possibly substituting among the shrimp species, especially those that are not currently imported to the United States. Thailand and China will continue to increase exports of shellfish, and Brazil is projected to expand its aquaculture production of shrimp and likely become a larger shrimp exporter than in the past. Imported salmon will increasingly originate from aquaculture farms because the wild-salmon catch may be reduced, in part, by regulations protecting various species. The ability of suppliers to meet currently held and yet-to-emerge food safety expectations will remain an issue.

CONCLUDING COMMENTS

Burke (1968) reported the pattern of changes in food consumption between 1910 and 1965. Food categories that showed a tendency to increase in importance were meat, poultry and fish, dairy products, and fruits and vegetables, while the consumption of potatoes and sweet potatoes and cereal products largely declined. But predictions based on the Engle function have been only partially confirmed by the observed consumption pattern. Variables other than disposable income, including the influence of food choices and many others, were listed in the section describing demand drivers. In recent years, among those factors, health considerations greatly increased in importance. The consumption of selected fresh and processed vegetables, fruits, nuts, and fish was affected by dissemination of research reports linking substances contained in these foods to confirmed or implied health benefits. For example, although Burke predicted the decline in sweet potato consumption, sweet potatoes have again become popular because of their high fiber, high vitamin A, and low energy content. Cultural factors, including ethnic and national traditions, beliefs, and customs, and their interaction with lifestyles will play an increasingly important role in food choices, including the consumption of fruits, vegetables, nuts, and seafood. Yet, in general, the effects of cultural factors on food demand and consumption patterns have received little attention in academia.

By 2030, the number of Americans 65 years old and older will double. Age is a major factor influencing the demand for convenience, including food choices and consumption. It is likely, therefore, that seafood consumption away from home will increase, while the selection of fresh fruit will be driven by the degree of effort it will take to prepare it for consumption and clean up afterward. Processed or partially processed fruits and vegetables will continue to be sought by the aging population. Elderly consumers already represent a substantial portion of the deli-section customers in supermarkets, buying prepared or partially prepared foods, including salads, sliced meats, and cooked seafood. Given the increasing cost of labor in the United States, supermarket chains will continue to seek suppliers able to deliver foods that will reduce handling and cooking at their stores. In their search, food distributors and chains will base their selections on the costs of purchase, suggesting that countries with low labor costs will be potential suppliers.

The Centers for Disease Control and Prevention recently issued a report indicating that Americans fail to meet the guidelines for the minimum consumption of fruits and vegetables (Anonymous, 2007). About 30% eat the recommended five fruit and vegetable servings per day (and this figure includes potatoes). The per capita seafood consumption has been increasing very slowly and is well below the recommended level as well. It appears that the increased imports of fruits, vegetables, and seafood marginally influence the per capita consumption but allow supermarkets and food service to compete for consumer food expenditures. The supermarket fruit and vegetable section influences the choice and frequency of visits by customers and, in part, determines store revenues. Imported produce has replaced domestically produced foods where production costs have forced American farmers to withdraw from production, has supplemented the offered choices, or has made certain foods available in the off-season. To induce Americans to eat more produce will require protection and maintenance of the integrity of the domestic food supply or consumers may lower their intake of fresh produce to avoid the risk of eating contaminated food.

Fresh produce importers have become acutely aware of the consequences of attributing a disease outbreak to imported fruits, vegetables, or seafood. Voluntary recalls after test results revealed the presence of *Salmonella* illustrate a proactive attitude on the part of some importers. For example, Castle Produce recalled Costa Rican cantaloupes in February 2007 because of *Salmonella* contamination (The Packer, 2007) and it was the second recall that month. A recall is less expensive and of much shorter duration than the consequences of a confirmed food-borne disease outbreak. However, voluntary recalls often occur after the contaminated imported fresh produce has already entered the supply chain in the United States. The threat of cross-contamination exists, and some of the produce might have been consumed, potentially infecting consumers. Although commendable, voluntary recalls should account for the speed at which the distribution system can withdraw contaminated produce, which should be as early and quickly as possible.

The potential barrier to nut consumption and import growth could be aflatoxin contamination. Edible nuts, except peanuts (which botanically are not nuts), are the only major food group that has not been consistently tested for aflatoxins. The recent decision by the European Union (EU) to draft the list of high-risk foods was stimulated by the large number of positive tests for aflatoxin in imported nuts, dried fruits, and spices (ElAmin, 2007). Although many of the countries exporting to the EU and some of the types of exported nuts are different from those shipped to the United States, aflatoxin contamination could disrupt the growth in edible nut imports and consumption.

Nutritionists and health officials will have to balance the risk of eating seafood, for example, salmon or tuna contaminated with heavy metals versus the intake of omega-3 fatty acids, because a serving of salmon contains the highest amount of omega-3 fatty acids among the top ten seafood types consumed in the United States. It appears that the message about the health benefits of eating seafood is heeded by the aging population. Also, the increase of Americans of Asian origin from 4% of the total population in 2000 to 5% in 2020 will likely contribute to increased seafood consumption (Ballenger and Blaylock, 2003).

The link between macroeconomic and other indexes and the level of food contamination risk is largely based on evidence that is fragmentary and anecdotal. However, the intention was to illustrate differences in the risk of microbial contamination for specific food products or categories. Changes in the level of risk will occur at the rate of changes in those indexes. By their very nature, the TI and the EFI as well as the PPP income change slowly over time, and changes in the country ranking from year to year should be considered with caution because perceptions that corruption interferes with conclusions regarding a foreign transaction may be affected by short-term phenomena, for example, economic conditions (Knack, 2006). Corruption also likely varies in degree among different sectors of administration within a country and can directly or indirectly affect the monitoring and implementation of food safety regulations or contractual obligations. Therefore, the cross-country and year-to-year comparisons can be revised periodically, but not annually. The intended use of the different indexes in this chapter is to enable the process of designing and enforcing a system for detecting microbial contamination and, under the best-case scenario, preventing contaminated food from reaching the distribution system.

The United States still imports only about 11% of its food consumption, a figure far lower than the one third of imported food consumed in the European Union. However, the U.S. food imports have been growing rapidly in recent years and the growth is likely to continue. A particular challenge to the food-safety-monitoring system is the uneven distribution of growth across different food categories, including especially hard-to-inspect seafood, fruits, and vegetables.

REFERENCES

Aguiar, M., and E. Hurst. 2006. Measuring trends in leisure: the allocation of time over five decades. Working paper no. 06-02, 61 p. Federal Reserve Bank of Boston, Boston, Mass.

Anonymous. 2007. Fruit and vegetable consumption among adults—United States, 2005. *Morb. Mortal. Wkly. Rep.* **56**:213–217.

Ballenger, N., and J. Blaylock. 2003. Changing U.S. demographics influence eating habits. *Amber Waves* 1(2):28–33. [Online.] http://www.ers.usda.gov/AmberWaves/April03/Features/ConsumerDrivenAg.htm. Accessed 27 March 2007.

Batres-Marquez, S. P., H. H. Jensen, and G. W. Brester. 2003. Salvadoran consumption of ethnic foods in the United States. *J. Food Distribution Res.* 34(2):1–15.

Broda, C., and D. Weinstein. 2005. Are we underestimating the gains from globalization for the United States? *Curr. Issues Econ. Finance* 11(4):1–7.

Burke, M. 1968. Consumption economics: a multidisciplinary approach. John Wiley & Sons, Inc., New York, N.Y.

Carew, R., W. J. Florkowski, and S. He. 2005. Contribution of health attributes, research investment and innovation to developments in the blueberry industry: a Canada-U.S. comparison. *Int. J. Fruit Sci.* **5**:95–116.

Carew, R., W. J. Florkowski, and E. G. Smith. 2006. Apple industry performance, intellectual property rights and innovation: a Canada-U.S. comparison. *Int. J. Fruit Sci.* **6**:93–116.

The Catfish Journal. 2007. Imported catfish grows at exposition. *Catfish J.* 21(8):1, 15.

CIA World Factbook. 2007. [Online.] https://www.cia.gov/library/publications/the-world-factbook/index.html. Accessed 17 March 2007.

The Economic Freedom Index. *The Wall Street Journal* and the Heritage Foundation. [Online.] http://en.wikipedia.org/wiki/Index_of_Economic_Freedom. Accessed 16 March 2007.

The Economic Freedom Index. *The Wall Street Journal* and the Heritage Foundation. [Online.] http://www.heritage.org/research/features/index/faq.cfm. Accessed 8 February 2008.

ElAmin, A. 2007. Rules would establish "high risk" list of imported foods. [Online.] http://www.foodproductiondaily.com/news/ng.asp?id=74652-imported-iran-rapid-alert. Accessed 2 May 2007.

Emerson, R. D., and F. Roka. 2002. Income distribution and farm labour markets, p. 137–149. *In* J. Findeis (ed.), *The Dynamics of Hired Farm Labour.* CABI Publishing, New York, N.Y.

Franz, N. 2006. Salmon—December 2006. [Online.] http://www.globefish.org/index.php?id=3476&easysitestatid=1480857738. Accessed 16 March 2007.

Huang, S., and F. Gale. 2006. China's rising fruit and vegetable exports challenge U.S. industries. U.S. Department of Agriculture, Economic Research Service, FTS-320-01, February. [Online.] http://www.ers.usda.gov/Publications/FTS/2006/02Feb/FTS32001/fts32001.pdf. Accessed 25 April 2007.

Jansen, H. G. 2006. Pro-poor economic growth in Latin America: obstacles and opportunities. *IFPRI Forum,* **September:**7.

Jerardo, A. 2003. Import share of U.S. food consumption stable at 11 percent. FAU-79-01, July. U.S. Department of Agriculture, Economic Research Service, Washington, D.C. [Online.] http://www.ers.usda.gov/publications/fau/july03/fau7901/fau7901.pdf. Accessed 5 March 2007.

Josupeit, H. 2006a. Tilapia Market Report—February 2007. [Online.] http:.//www.globefish.org/index.php?id=3614. Accessed 16 March 2007.

Josupeit, H. 2006b. Tuna Market Report—November 2006. [Online.] http://www.globefish.org/index.php?id=3376. Accessed 16 March 2007.

Knack, S. 2006. Measuring corruption: a critique of cross-country indicators. *World Bank Res. Dig.* 1(2):6.

Lewis., E. G. 2007. The impact of immigration on American workers and businesses. A paper presented at the SAAS Annual Conference, 3–6 February, Mobile, Alabama.

Lin, B-H. 2004. Fruit and vegetable consumption— looking ahead to 2020. U.S. Department of Agriculture, Economic Research Service, Agriculture Information Bulletin 792-7, October. [Online.] http://151.121.68.30/publications/aib792/aib792-7/aib792-7.pdf. Accessed 28 March 2007.

Lin, B-H., J. N. Variyam, J. Allshouse, and J. Cromartie. 2003. Food and agricultural commodity consumption in the United States: looking ahead to 2020. U.S. Department of Agriculture, Economic Research Service, Economic Report no. 820, February. [Online.] http://www.ers.usda.gov/publications/aer820/aer820.pdf. Accessed 28 March 2007.

Lopez, J. 2007. Shrimp market report: March 2007—US. [Online.] http://www.globefish.org/index.php?id=3629. Accessed 15 March 2007.

Mehta, K., S. M. Gabbard, V. Barrat, M. Lewis, D. Carroll, and R. Mines. 2000. Findings from National Agricultural Worker Survey (NAWS) 1997–1998: A demographic and employment profile of U.S. farm workers. Research Report no. 8, U.S. Department of Labor, March. As cited by Napsintuwong and Emerson in Labor Substitutability in Labor Intensive Agriculture and Technological Change in the Presence of Foreign Labor, Selected Paper presented at the AAEA Annual Meeting, Denver, Colorado, 1–4 August 2004.

Messick, R. E., and R. Kleinfeld. 2001. Writing and effective anticorruption Llw. *Transition Newslett.* **12**(4):10–11.

Msangi, S., and M. W. Rosegrant. 2005. Fish to 2020: supply and demand in changing global markets. Presentation at Global Aquaculture Meeting, 20 June 2005, Seattle, Wash. [Online.] http://www.lib.noaa.gov/docaqua/noaa_matrix_program_reports/03_siwa_msangi.pdf Accessed 27 March 2007.

National Marine Fisheries Service (NMFS). 2007a. Per capita consumption. [Online.] http://www.st.nmfs.gov/st1/fus/fus05/08_perita2005.pdf.

National Marine Fisheries Service (NMFS). 2007b. Top 10 U.S. consumption by species chart. [Online.] http://www.aboutseafood.com/media/top_10.cfm. Accessed 20 December 2007.

The Packer. 2007. In Brief. March 5, p. A4.

Park, T. A., and W. J. Florkowski. 1999. Demand and quality uncertainty in pecan purchasing decisions. *J. Agric. Appl. Econ.* **31**(1):29–39.

Transparency International. 2004 Annual Report. [Online.] http://www.transparency.org. Accessed 16 March 2007.

Trejo, S. J. 1998. Immigrant Participation in Low-Wage Labor Markets, University of California Santa Barbara. Mimeo, December.

U.S. Department of Agriculture (USDA). 2007. Outlook for U.S. agricultural trade. AES-53. March 1, 2007. [Online.] http://www.fas.usda.gov/cmp/outlook/2007/Mar-07/AES-03-01-2007.pdf. Accessed 19 March 2007.

U.S. Department of Agriculture, Economic Research Service (USDA-ERS). 2007. Economics of foodborne disease. [Online.] http://www.ers.usda.gov/Briefing/FoodborneDisease/. Accessed 19 March 2007.

U.S. Department of Agriculture, Economic Research Service (USDA-ERS). 2006a. Food CPI, prices and expenditures: food expenditures by families and individuals as a share of disposable personal income. [Online.] http://www.ers.usda.gov/briefing/CPIFoodAndExpenditures/Data/table7.htm. Last update 9 June 2006. Accessed 1 February 2007.

U.S. Department of Agriculture, Economic Research Service (USDA-ERS). 2006b. Food CPI, prices and expenditures: food service as a share of food expenditures. [Online.] http://www.ers.usda.gov/Briefing/CPIFoodAndExpenditures/Data/table12.htm. Updated 15 September 2006. Accessed 1 February 2007.

U.S. Department of Agriculture-Economic Research Service (USDA-ERS). 2005. Commodity highlight: fresh-market apples. *Fruit and Tree Nuts Outlook.* FTS-315, March. [Online.] http://www.ers.usda.gov/publications/fts/mar05/FTS315.pdf. Accessed 23 April 2007.

U.S. Department of Agriculture, Economic Research Service (USDA-ERS). 2003. Pineapple production concentrated in tropical regions of the world. *Fruit and Tree Nuts Outlook.* FTS-307, November. [Online.] http://www.ers.usda.gov/publications/fts/nov03/fts307.pdf. Accessed 25 April 2007.

U.S. Department of Commerce (USDC). 2006. U.S. Census Bureau. Average population per household and family: 1940 to present. [Online.] http://www.census.gov/population/socdemo/hh-fam/hh6.xls. Internet release 21 September 2006. Accessed 30 January 2007.

U.S. Department of Commerce (USDC). 2005. U.S. Census Bureau.: Income. [Online.] http://www.census.gov/hhes/www/income/histinc/f06ar.htm. Last revised 20 December 2005. Accessed 30 January 2007.

U.S. Department of Labor (USDL). 2006. Indexes of Hourly Compensation Costs. Last modified date 30 November 2006. [Online.] http://www.bls.gov/news.release/ichcc.t01.htm. Accessed 6 March 2007.

U.S. Import Edition, vol. 11, no. 4. [Online.] http://www.SeafoodReport.com. Accessed 14 March 2007.

Imported Foods: Microbiological Issues and Challenges
Edited by Michael P. Doyle and Marilyn C. Erickson
© 2008 ASM Press, Washington, DC

Food Safety Regulations Applicable to Imported Foods

2

Neal D. Fortin

INTRODUCTION

Overview of the Regulatory System

Regulatory oversight of imported foods is the focus of this chapter and is based on legislation that has, in part, been in effect for up to 100 years. Despite this long history, recent changes to bolster the authority of the agencies responsible for regulating imported food have been proposed. Hence, it is important to recognize that information presented in this chapter pertains to food safety regulations that were in effect as of October 2007. Proposed changes in regulatory activities that are under consideration will be addressed in chapter 9. (Portions of this chapter are derived from Fortin, *Food Regulation: Law, Science, Policy, and Practice* [in press] and are used with the permission of the publisher.)

U.S. law requires that all imported foods meet the same food safety standards as foods produced in the United States. The Federal Food, Drug, and Cosmetic Act (Pub. L. No. 75-717, 52 Stat. 1040 (1938), as amended, 21 U.S.C. sections 301–397 (2000) [FDCA]) and other laws that are designed to protect consumers' health and safety apply equally to domestic and imported products. For instance, imported foods must be pure, wholesome, safe to eat, and produced under sanitary conditions. Furthermore, foods must bear the same informative and truthful labeling in English.

The Food and Drug Administration (FDA) and the U.S. Department of Agriculture (USDA) Food Safety Inspection Service (FSIS) share primary responsibility for ensuring the safety of food imported into the United States.

NEAL D. FORTIN, Professor and Director, Institute for Food Laws and Regulations, Michigan State University, East Lansing, MI 48824.

FSIS has responsibility over meat, poultry, and some egg products. FDA regulates all other foods.

FSIS inspects each shipment of meat, poultry, and egg products imported to the United States. In addition, FSIS is required to determine that the exporting country has a food safety inspection system for the products that is equivalent to the U.S. system. In contrast, FDA lacks the statutory authority to impose an equivalency requirement for importation of FDA-regulated foods. FDA generally must rely on inspections at the U.S. ports of entry to determine the safety of the imported foods.

This simple overview, however, masks a relatively complex set of interconnected regulations enforced by a number of different agencies. There are a number of redundancies in the import procedures because of overlapping requirements for specific food products (General Accounting Office [GAO], 2005). Moreover, the basic structure of import regulation for most food was put in place in 1906 and the legal authorities and resources have not kept pace with the changes due to the globalization of the food supply and increased imports.

The Major U.S. Federal Agencies

To gain an understanding of the food safety regulatory system for imported foods, first one must sort through the alphabet soup of agencies that enforce the regulations. Eight federal agencies play a major role in the regulation of imported foods:

- FDA, U.S. Food and Drug Administration
- USDA, U.S. Department of Agriculture
  - APHIS, Animal and Plant Health Inspection Service
  - FSIS, Food Safety Inspection Service
  - NCIE, National Center for Import and Export (NCIE) Veterinary Services
- CBP (Customs), Bureau of Customs and Border Protection
- EPA, Environmental Protection Agency
- TTB, Alcohol and Tobacco Tax and Trade Bureau
- NMFS, National Marine Fisheries Service

Both FSIS and FDA depend on working closely with the Bureau of Customs and Border Protection (CBP or "Customs") (FDA, 1979). Customs notifies FSIS and FDA of imported foods for the agencies' review. Customs holds imported food from commerce until the release by FSIS or FDA. Another major responsibility of Customs is to administer the Tariff Act of 1930 and assess and collect all duties, taxes, and fees on imported merchandise. The agency also administers and reviews import entry forms.

State agencies also play a role in import regulations. Imported product must conform to all the requirements in each of the 50 states where the product is sold—in addition to the federal laws of the United States. Fifty plus sets of differing regulations could be an immense burden to commerce, but, in general, most state requirements are consistent with the federal requirements. One notable exception is California's Prop 65, which requires special warning statements on many products. Other exceptions exist but generally are smaller in scope and apply to a limited category of foods. For example, Michigan law requires a "last date of sale" on certain perishable foods, while the federal law is silent in this area.

IMPORT PROCESS

Foods may be imported into the United States if they meet the same standards as those foods that are produced domestically. However, imported food faces significant procedural and legal hurdles that are higher than domestic food products face. In particular, the standard for import denial is the product appears to be adulterated or misbranded, while domestic goods cannot be condemned unless they actually are shown to be adulterated or misbranded (FDCA section 801; 21 U.S.C. section 381). This creates a daunting standard of proof for an importer who wishes to challenge the FDA's determination (see, e.g., *Goodwin v. U.S.*).

In addition, importers face fewer constitutional protections than owners of food already in U.S. commerce (*Continental Seafoods, Inc. v. Schweiker*, 1982). For example, condemnation of a domestic food for adulteration deprives the owner of value, but import denial is not the taking of the importer's property (*Meserey v. U.S.*, 1977). There is no constitutional right to import goods into the United States, and due process protections apply only after the food enters U.S. commerce. Moreover, the courts give FDA broad discretion in the measurement of defects in imported foods (*Caribbean Produce Exchange, Inc. v. Secr. H.H.S.*, 1989).

FDA

FDA regulates the importation of most foods other than meat, poultry, and some egg products. FDA's authority derives from section 801 of the Food Drug and Cosmetic Act (FDCA), which authorizes FDA examination of foods, drugs, cosmetics, and medical devices offered for entry into the United States (21 U.S.C. 381). This authority was largely put in place in 1906, when Congress passed the Pure Food and Drug Act of 1906. The provisions were carried over when the Federal Food, Drug, and Cosmetic Act was enacted in 1938.

Basic import procedure

Within 5 working days of the date of arrival of a shipment of food at a port of entry, the importer must file entry documents with U.S. Customs (FDA, 1996). FDA is notified of the entry documents and reviews the importer's entry documents to determine whether a physical examination should be made or a sample taken for analysis. FDA's decision on whether to collect a sample is based on the nature of the product, FDA priorities, and the history of the commodity (FDA, 1996).

If the decision is made not to collect a sample, FDA sends a "May Proceed Notice" to Customs and the importer, and the shipment is released as far as FDA is concerned. If FDA sends a "Notice of Sampling" to Customs and the importer, the shipment must be held intact pending further notice, but the importer may move the shipment from the dock to another port or warehouse.

This system, where the importer rather than FDA retains custody over shipments, has been criticized for allowing shipments that failed to meet U.S. safety standards to be distributed in domestic commerce (GAO, 1998a). Importers in some cases may have been able to provide substitutes for food targeted for inspection.

If the sample is found in compliance with requirements, FDA sends a "Release Notice" to Customs and the importer. If the sample appears to be in violation, FDA sends Customs and the importer a "Notice of Detention and Hearing."

Prior notice of import

Prior notice is notification to the FDA that an article of food or animal feed is being imported or offered for import into the United States in advance of the arrival of the article of food at the U.S. border (68 Fed. Reg. 58974 [Oct. 10, 2003]) (see also FDA [May 2004] Guidance for Industry: Prior Notice of Imported Food Questions and Answers, 2nd Ed. Available at: http://www.cfsan.fda.gov/~pn/pnqagui2.html. Last visited 27 December 2006). The Bioterrorism Act of 2002 (Public Health Security and Bioterrorism Preparedness and Response Act of 2002 [the Bioterrorism Act] [Public Law 107-188]) added a requirement to the FDCA that FDA receive prior notice of food imported into the United States [FDCA section 801(m) (21 U.S.C. 381(m)], which was added by section 307 of the Bioterrorism Act). On October 10, 2003, FDA published an interim final rule requiring submission to FDA of prior notice of food and animal feed that is imported or offered for import into the United States (68 Fed. Regist. 58974 [10 October 2003]).

Because of the broad definition of "food" under the FDCA, the FDA's prior notice requirement applies to some products also regulated by other agencies. For example, alcoholic beverages regulated by TTB must still comply with FDA prior notice requirements. Live food animals that are subject to border inspections by APHIS are also subject to FDA's prior notice requirements (live food animals do not fall within the exclusive jurisdiction of USDA under the Federal Meat Inspection Act or Poultry Products Inspection Act, thus FDA and APHIS may both have jurisdiction over live animals) (68 Fed. Regist. 58974 at p. 58991 [10 October 2003]).

The requirement for prior notice to the FDA does not alter the role of another agency, such as APHIS or TTB, or the requirements relating to that agency (68 Fed. Regist. 58974 at p. 58991 [10 October 2003]). However, food under the exclusive jurisdiction of the USDA at the time of importation is excluded from the prior notice requirement.

Import food facility registration

The Bioterrorism Act also requires domestic and foreign facilities that manufacture, process, pack, or hold food for human or animal consumption in the United States to register with the FDA. Farms, fishing vessels not engaged in processing, and facilities regulated exclusively throughout the entire facility by the USDA are exempt from registration. Registration may be done by paper form, but FDA encourages electronic registration, which is available on the FDA's Web site (see http://www.fda.gov/furls).

Additional forms for certain canned foods, milk, cream, and infant formula

In addition to the required entry forms, import registration, and prior notice of import, certain food products require additional specific information be provided to FDA. Firms must register and file processing information before shipping any low-acid canned food or acidified low-acid canned food into the United States (21 Code of Federal Regulations [C.F.R.] sections 108.25 and 108.35). This information must be provided to the FDA for each applicable product at the time of importation. In addition, the Federal Import Milk Act requires a permit for milk and cream imported into the United States (21 U.S.C. sections 141–149). These permits and registrations are in addition to the general registration and prior notice requirements.

Infant formula, in addition to meeting the laws and regulations governing foods in general, must meet additional statutory and regulatory requirements. The specific infant formula requirements are found in FDCA section 412 and 21 C.F.R. sections 106 and 107. In particular, all formulas marketed in the United States must meet federal nutrient requirements and infant formula manufacturers must notify the FDA before marketing a new formula. This is in addition to other notification requirements. If an infant formula manufacturer does not provide the information required in the notification for a new or reformulated infant formula, the formula is defined as adulterated under FDCA section 412(a)(1). These more stringent requirements were considered necessary because infant formula is often used as the sole source of nutrition by a vulnerable population during a critical period of growth and development.

Although FDA's statutory authority is largely limited to inspections and tests of imported foods at the U.S. port of entry, with low-acid and acidified canned foods and infant formula, FDA may request that foreign exporting firms grant FDA inspectors access to their plants. Nonetheless, FDA conducts few foreign plant inspections. For example, there are almost 190,000 foreign food firms exporting food to the United States, yet the FDA inspected fewer than 100 firms in fiscal year 2007 (GAO, 2008).

When a violation is found

The FDA may refuse entry of an import shipment after a paperwork inspection and a physical examination. The FDCA authorizes the FDA to detain a regulated product that appears to be out of compliance with the FDCA. The FDA district office will then issue a "Notice of FDA Action," which identifies the nature of the violation to the owner or consignee of the goods.

The owner (or consignee) is entitled to an informal hearing regarding the admissibility of the goods; however, the hearing is less than FDA's full regulatory hearing (21 C.F.R. sections 1.94 and 16.5(a)(2)). The hearing is the importer's only opportunity to present a defense of the importation or to present evidence as to how the shipment may be made eligible for entry.

The importer faces a steep burden of proof at such a hearing. In particular, the importer cannot demand that the FDA prove the source of contamination, and FDA only has to prove that a product "appears adulterated" (21 U.S.C. section 381). This language in the FDCA indicates "Congress' intent to forego formal procedural requirements" for imports (*Seabrook Intl. Foods, Inc. v. Harris*, 1980). Typically, the courts grant FDA broad deference and discretion in measuring and examining defects in imported foods (see, e.g., *Caribbean Produce Exchange, Inc. v. Secr. of HHS*, 1989).

If the owner fails to submit evidence that the product is in compliance or fails to submit a plan to bring the product into compliance, FDA issues a second "Notice of FDA Action," which refuses admission to the goods (Fig. 1). The goods must then be exported or destroyed within 90 days (FDA, 2002). Reexportation is within the discretion of FDA (21 U.S.C. section 334(d)(1)).

Request for authorization to relabel or perform other acts

The importer of detained goods may propose a manner in which detained food can be brought into compliance with the FDCA or be removed from coverage under the FDCA (that is, rendered other than a food, drug, device, or cosmetic). The FDA may authorize relabeling or other action based on a timely submission of an appropriate completed request and the execution of sufficient bond [21 U.S.C. section 381(b) and 21 C.F.R. section 1.95]. The FDA notifies the importer of the approval or disapproval of the application to relabel or recondition. The FDA can charge the importer for the costs of supervision of the relabeling or reconditioning (21 C.F.R. section 1.99). When approved, FDA will state the conditions to be fulfilled, and the time limit within which to fulfill (21 C.F.R. section 1.96).

Inspection after reconditioning or relabeling

After completion of relabeling or reconditioning, the importer provides FDA with notification of completion. At this point the FDA may conduct a follow-up inspection, sampling, or both to determine compliance. The FDA may also accept the statement from the importer and conduct no follow-up.

If the relabeling or reconditioning has been properly fulfilled, the FDA will notify the owner or consignee that the admissible portion is no longer subject to detention or refusal of admission. This notice is usually identified as "Originally Detained and Now Released." Where a nonadmissible portion remains (rejects), that portion must be destroyed or reexported under FDA or Customs supervision. A "Notice of Refusal of Admission" is issued for the rejected portion.

If the relabeling or reconditioning has not been successfully fulfilled, the FDA generally will not authorize a second relabeling or reconditioning unless the request includes an adjustment from the original method, and the applicant offers reasonable assurance that the second attempt will be successful. If an article is refused admission, such article must be destroyed or exported under Customs' supervision, generally within 90 days of receiving the Notice of Refusal [21 U.S.C. section801(a)].

Good agricultural practices (GAPs)

In 1998, FDA issued the "Guide to Minimize Microbial Food Safety Hazards for Fruits and Vegetables." This guide recommends good agricultural practices (GAPs) and good manufacturing practices (GMPs) that growers, packers, and shippers should perform to address common risk factors and reduce the food safety hazards potentially associated with fresh produce (FDA, 1998a).

Although the GAPs contain important guidelines for food safety practices, they are not binding on growers, packers, or shippers of food. Nonetheless, GAPs remain an important contribution to the control of food safety in

United States Food and Drug Administration
Los Angeles District Office

Notice of FDA Action
Entry Number: 112-9861457-6 Notice Number: 2 November 6, 1996

Filer:
FBN Freight Services Attention: George
500 Canal St.
New Orleans LA 70130
Port of Entry: 2704, Los Angeles,

Carrier: NOL RUBY

Entry Date: November 2, 1996
Arrival Date: November 4, 1996
Importer of Record: Shipley'S Donut Shop Inc., Lafayette, LA
Consignee:
a: Shipley'S Donut Shop Inc., Lafayette, LA
b: Specialty Commodities Inc. Fargo ND

HOLD DESIGNATED

Documents Required and Notify FDA of Availability

Summary of Current Status of Individual Lines

@ LINE

ACS/FDA Product Description Quantity Current Status

* a 001/001 PINEAPPLE, DEHYDRATED 500 CT RELEASED 11-6-96

@ LINE

ACS/FDA Product Description Quantity Current Status

* a 002/001 DEHYDRATED GINGER SLICES 10 KG Product Collected by FDA

11-06-96

@LINE

ACS/FDA Product Description Quantity Current Status

* b 003/001 PAPAYA, DEHYDRATED 10 KG Detained 11-06-96

* = Status change since the previous notice. Read carefully the sections which follow for important information regarding these lines.

Figure 1 Example of "Second Notice of FDA Action."

@ = Consignee id

Notice of FDA Action Notice Number: 2

Entry Number: 112-9861457-6 Page: 2

FDA will not request redelivery for examination or sampling, if the products not released by FDA are moved, following USCS conditional release to a location within the local metropolitan area or to a location approved by the FDA office at the number below.

All products in this entry not listed above may proceed without FDA examination. This notice does not constitute assurance the products involved comply with provisions of the Food, Drug, and Cosmetic Act or other related acts, and does not preclude action should the products later be found violative.

Please provide documentation concerning all products in this entry to the FDA office below. Include the USCS document (eg. CF-3461 or CF-7501) and commercial invoice for these products, annotated to show the ACS/FDA line numbers sent electronically.

Also, advise FDA upon actual availability, and include date, location, and warehouse control number, where applicable, for all lines in this entry.

Jennifer A Thomas, Inspector

U.S. Food & Drug Administration (213-555-1212)

2nd and Chestnut Streets (HFR-MA100)

Philadelphia, PA 19106

DETENTION WITHOUT EXAMINATION

The following products are subject to refusal pursuant to the Federal Food Drug and Cosmetic Act (FD&CA), Public Health Service Act (PHSA), or other related acts in that they appear to be adulterated, misbranded or otherwise in violation as indicated below:

LINE

ACS/FDA Product Description Respond By

003/001 Product: PAPAYA, DEHYDRATED November 26, 1996

FD&CA Section 402(a)(1), 801(a)(3); ADULTERATION

The article appears to be held in a container containing a poisonous or deleterious substance which may render it injurious to health.

FD&CA Section 402(a)(2)(B), 801(a)(3); ADULTERATION

The article appears to be a raw agricultural commodity that bears or contains a pesticide chemical which is unsafe within the meaning of Section 408(a). The article appears to contain quinalphos.

U. S. Food & Drug Administration

Notice of FDA Action Notice Number: 2

Entry Number: 112-9861457-6 Page: 3

Figure 1 (*continued*)

You have the right to provide oral or written testimony, to the Food & Drug Administration, regarding the admissibility of the article(s) or the manner in which the article(s) can be brought into compliance. This testimony must be provided to FDA on or before the dates shown above.

SAMPLES COLLECTED

LINE

ACS/FDA Product Description Est. Cost

001/001 PINEAPPLE, DEHYDRATED $ 15.00

Sample: 10 KG Collected 1 KG from each of 10 cartons

LINE

ACS/FDA Product Description Est. Cost

002/001 DEHYDRATED GINGER SLICES $ .23

Sample: .1 KG Collected approximately 4 ounces from one carton.

LINES RELEASED

LINE

ACS/FDA Product Description

001/001 PINEAPPLE, DEHYDRATED

These products are released. This notice does not constitute assurance that the product released complies with all provisions of the Food, Drug, and Cosmetic Act, or other related Acts, and does not preclude action should the product later be found violative.

Notice Prepared by: Thomas J DiNunzio (QA5)

U.S. Food & Drug Administration

Figure 1 *(continued)*

imports. The FDA's actions can leverage its authority in ways that gain voluntary compliance with the GAPs. In particular, the FDA can detain or ban shipments from growers or countries. For example, in 2002, the FDA banned all cantaloupe imports from Mexico (Calvin, 2003). This action may explain some of the growth in the use of third-party audits and certification for GAPs. In addition, market forces can create a prophylactic factor that encourages adoption of better food safety measures, such as GAPs.

U.S. Department of Agriculture (USDA)
Food Safety Inspection Service (FSIS)
USDA-FSIS regulates the importation of meats and poultry and some egg products. The duty to inspect all commercial shipments of meat and poultry products entering the united States has been delegated to the USDA under the authority of the Federal Meat Inspection Act of 1958 (FMIA) and

the Agricultural Marketing Agreement of 1937. The applicable USDA regulations appear generally at Title 9 of the Code of Federal Regulations.

Before foreign firms can export meat or poultry into the United States, FSIS must have determined that the exporting country has a meat or poultry food safety system that is equivalent to that in the United States. When the FSIS receives an application, the agency compares the foreign inspection system with the measures FSIS applies domestically. If the FSIS determines that the foreign food regulatory system documentation meets all U.S. import requirements in the same or an equivalent manner, and provides the same level of public health protection attained in the United States, the FSIS conducts an on-site audit of the entire foreign meat or poultry food regulatory system (or both). If a country completes these steps satisfactorily, the FSIS publishes a proposed regulation that would add the country to FSIS' list of eligible import countries. FSIS must collect public comments on this proposed regulation and consider the comments before making a final decision as to whether the country will be eligible to import meat, poultry, or egg products into the United States.

The time from application to FSIS to the completion of an initial equivalence approval process normally requires 3 to 5 years. In October 2007, a total of 33 countries were eligible to import meat and poultry products to the United States (Table 1).

After initial approval, the FSIS utilizes a three-part process to verify that foreign meat or poultry food regulatory systems continue to be equivalent to those in the United States. First, the FSIS reviews documents, such as the laws, regulations, and implementing policies of a foreign food regulatory system, to ensure that the infrastructure remains in place. Next, the FSIS conducts on-site food regulatory system audits at least annually in every country that exports meat or poultry products to the United States. Third, FSIS's continuous port-of-entry reinspection of products shipped from exporting countries provides evidence of how the foreign inspection systems are functioning.

In contrast to the FDA, FSIS inspectors visually check every imported shipment of foods under their jurisdiction at FSIS-approved import inspection stations. Most of these checks are for correct documentation and labeling. FSIS conducts more complete inspections and tests on a portion of the imported shipments to verify the effectiveness of the foreign food safety system. In 1997, 20% received more complete inspections (GAO, 1998a), whereas in the first three quarters of FY 2007, 11% received more complete inspections (FSIS, 2007b). FSIS uses the term "reinspection" for its imported product inspections because the products have been previously inspected and passed by the importing country's inspection system.

Table 1 Eligible foreign meat and poultry establishments (FSIS, 2007a)

Eligible country	Type of product
Argentina	Meat
Australia	Meat, poultry (ratites only)
Belgium	Meat
Brazil	Meat
Canada	Meat, poultry; egg products
Chile	Meat
China	Poultry
Costa Rica	Meat
Croatia	Meat
Czech Republic	Meat
Denmark	Meat
Finland	Meat
France	Meat, poultry
Germany	Meat
Honduras	Meat
Hungary	Meat
Iceland	Meat
Ireland	Meat
Israel	Poultry
Italy	Meat
Japan	Meat
Mexico	Meat, poultry[a]
The Netherlands	Meat
New Zealand	Meat, poultry (ratites only)
Nicaragua	Meat
Northern Ireland	Meat
Poland	Meat
Romania	Meat
San Marino	Meat
Spain	Meat
Sweden	Meat
United Kingdom	Meat, poultry
Uruguay	Meat

[a]Mexico is approved to export only processed poultry products slaughtered under federal inspection in the United States or in a country eligible to export slaughtered poultry to the United States.

In the same manner as importers of FDA-regulated products, importers of FSIS-regulated products must file an import notice and a bond with Customs within 5 days of the date that a shipment arrives at a port of entry. Unlike FDA law, which allows shipments to be moved out of FDA control, importers of FSIS-regulated food must hold their shipments at FSIS-registered warehouses for FSIS inspection until these shipments are released or refused entry.

However, before an importer may bring FSIS-regulated products into the United States, the importer must be certain its country has been accepted by FSIS to sell meat, poultry, or egg products in the United States. FSIS must determine that a country's federal inspection system is equivalent to that of the United States. FSIS does not conduct food inspections in foreign countries, nor does it verify that individual foreign establishments are qualified to export to the United States. The FSIS relies on its determination that a country has an equivalent food regulatory system that carries out appropriate inspection. A foreign establishment must obtain certification from its country's chief inspection official, who will certify to the FSIS which establishments in the country meet FSIS' import requirements.

Animal and Plant Health Inspection Service (APHIS)

USDA-APHIS is charged with protecting U.S. agricultural health, among many other responsibilities, such as regulating genetically engineered organisms and administering the Animal Welfare Act. To accomplish its mission, one role of APHIS and the USDA Veterinary Services (VS) is to regulate the importation of animals and animal-derived materials to ensure that exotic animal and poultry diseases are not introduced into the United States. For example, APHIS works to prevent entry of foot-and-mouth disease or avian influenza. Each meat, poultry, and egg product shipment that enters the United States first falls under the control of Customs and APHIS. If no pest or disease of concern is detected or raised by the documentation, APHIS transfers control of the products to FSIS for visual inspections.

Customs (CPB)

All FDA- and USDA-regulated products imported into the United States must meet the Bureau of Customs and Border Protection (CBP or "Customs") requirements in addition to FDA and USDA requirements. The major responsibility of Customs is to administer the Tariff Act of 1930 as amended. Primary duties include assessment and collection of all duties, taxes, and fees on imported merchandise; administration and review of import entry forms; the enforcement of Customs-related laws; and administration of certain navigation laws and treaties.

There is a working agreement among FDA, USDA, and Customs for cooperative enforcement. Products nonconforming with FDA or USDA requirements will be seized by Customs and released only after the agency receives written approval from FDA or USDA, as applicable. In general, the

FDA or USDA identify the violative food, but the refusal of admission, and subsequent reexportation or destruction of the food, is carried out under the direction of Customs. In some cases, actual supervision of destruction of violative food or the supervision of reconditioning may be conducted by FDA or FSIS personnel under a regional agreement. For example, where a port of entry is close to an FDA office, supervision is normally exercised by the FDA. At remote ports, supervision is normally exercised by Customs (FDA, 2002).

The FDA has an electronic notification entry system, the Operational and Administrative System for Import Support (OASIS). When Customs receives electronic notifications of a food shipment entry, which are sent to the FDA electronically via OASIS, the FDA uses OASIS to electronically screen entries against criteria developed by the FDA.

Articles offered for import into the United States (entries) that have a value greater than $1,250 are considered by Customs to be "formal" entries. One of the more important requirements for formal entries is the requirement for a bond. Under a formal entry bond, imported articles may be unconditionally released to importers, pending a determination of the admissibility (and amount of duty to be paid). The bond requires importers to redeliver the articles to Customs, upon demand of Customs at any time. For example, Customs might demand redelivery of a food to allow FDA sampling or for reexportation following refusal of admission. If the importer fails to redeliver the goods, Customs may institute proceedings to collect the liquidated damages provided for in the bond [see 19 C.F.R. 113.62(k) and 21 C.F.R. 1.97].

Under FDA law, importers generally maintain possession of the imported food before the FDA releases them. With perishable foods, the shipment may begin entry into domestic commerce. This system has been criticized on a number of grounds (GAO, 1998a). Not all foods sampled, and later found violative, are returned by importers to Customs. In addition, even when food is returned, Customs does not always witness and verify that violative food is properly disposed of; for instance, a landfill receipt may suffice. Customs also does not verify whether there has been a substitution when product is reexported from the United States instead of destroyed. Finally, forfeiture of the bond is not always effective deterrence to ensure return of the food, either because the value of the food may exceed the bond or because full damages often are not collected (GAO, 1998a).

Environmental Protection Agency (EPA)

The EPA is not directly involved in the regulation of imported food, but imported food must meet the same pesticide residue standards as domestic product. The law directs the EPA to set limits on the pesticide residue remaining on food such that there is a reasonable certainty of no harm. These pesticide residue limits are called tolerances (some countries use the term "maximum residue limits" or MRLs). EPA sets the tolerances in the Code of Federal Regulations within Title 40 C.F.R. Part 180. (The tolerance information is codified within 40 C.F.R. 180. However, for information on new tolerances or changes to tolerances not yet codified, search the EPA's Web site http://www.epa.gov/pesticides/search.htm or the Federal Register (available at: http://www.gpoaccess.gov/fr/index.html.)

The pesticide tolerances apply equally to imported and domestically produced food. (For more information on pesticide tolerances, see U.S. Environmental Protection Agency, Pesticides and food: what the pesticide residue limits are on food, available at: http://www.epa.gov/pesticides/food/viewtols.htm. Last accessed 3 January 2007). These tolerances are enforced by the USDA and FDA. The USDA enforces tolerances established for meat, poultry, and some egg products. The FDA enforces the tolerances established for other foods.

Alcohol and Tobacco Tax and Trade Bureau (TTB)

Importers seeking to import alcoholic beverages into the United States must meet the requirements of the Federal Alcohol Administration Act enforced by the Alcohol and Tobacco Tax and Trade Bureau (TTB) (For more information, see http://www.ttb.gov/index.shtml. Last accessed 7 February, 2007.) In particular, an importer must obtain the appropriate TTB-issued permit to import alcoholic products. Importers must maintain and staff a business office in the United States. In addition, the importer must have a TTB-issued certificate of label approval (COLA). Finally, the importer must meet all state and local requirements, which may be in addition to federal requirements.

Alcoholic beverages are also defined as "food" under other statutes, so alcoholic beverages must also meet those additional general requirements. For example, the importer must ensure that the producer of the alcoholic beverage is registered with the FDA and provide FDA with advance notification of an importation.

National Oceanic and Atmospheric Administration (NOAA)

More than 80% of the seafood that Americans consume is imported (GAO, 2007). Seafood falls under the regulatory oversight of the FDA. However, voluntary inspection programs within the National Oceanic and Atmospheric Administration (NOAA) of the U.S. Department of Commerce provide important support for FDA's regulatory role. Administered through the 1946

Agricultural Marketing Act, these programs include establishment, sanitation inspection, process and product inspection, product grading, product lot inspection, laboratory analyses, training, and consultation (USDC, 2007). While this is not strictly speaking an import program, the service is provided in foreign countries as well as the United States.

Products that are inspected and meet the requirements under the program can bear one of the agency's official marks, such as U.S. Grade A, Processed under Federal Inspection (PUFI), and lot inspection marks. The program is available for all edible products, ranging from whole fish to formulated products, as well as fishmeal products for animal foods.

The NOAA also plays a role in seafood imports through its division of the National Marine Fisheries Service (NMFS). The NMFS is responsible for the management, conservation, and protection of living marine resources. The agency assesses and predicts the status of fish stocks, ensures compliance with fisheries regulations, and works to reduce wasteful fishing practices and to prevent lost economic potential associated with overfishing, declining species, and degraded habitats. In these roles, NMFS may put restrictions on the import of certain marine species. For instance, to implement recommendations of the International Commission for the Conservation of Atlantic Tunas, NMFS has banned the import of undersized Atlantic swordfish and extended dealer permitting and reporting requirements to swordfish importers (50 C.F.R. section 635.46).

Codex Alimentarius

The Codex Alimentarius is Latin for food book or food code. The Codex is a set of international food standards, codes of practice, guidelines, and recommendations. The importance of Codex for international food trade has grown in recent years to become the world's reference point for international food trade. The Agreement on the Application of Sanitary and Phytosanitary Measures (the SPS Agreement), for example, cites Codex standards as the international measures for trade in food. The Technical Barriers to Trade Agreement (TBT) similarly encourages use of international food standards. As such, the Codex standards have become the benchmark against which national food regulations are evaluated within the legal parameters of the World Trade Organization (WTO). Essentially, countries must base their sanitary and phytosanitary measures on Codex standards unless a scientific basis exists for a greater level of protection (FAO/WHO, 2005).

In addition, many countries adopt Codex standards into their national food laws and regulations. Codex has become not just a means to harmonize the international community but also to ensure global implementation of food safety standards and codes of practices. Thus, Codex is both a means to assist countries in producing safe food and a means for importing countries to ensure that appropriate standards have been applied.

CHALLENGES FACING IMPORT REGULATION

The emergence of bovine spongiform encephalopathy (BSE) as a human health and food safety issue in 1996 prompted U.S. regulatory agencies to establish additional policies to ensure that contaminated product did not enter the marketplace. These policies have been continuously reviewed and updated as new issues subsequently have arisen. One of the earliest controls put in place was the ban of live cattle or beef and beef products exported from the United Kingdom where a high incidence of cases originated. Later, this ban extended to any country with confirmed cases of BSE (FAO/WHO, 2002). Countries currently listed on the FDA Web site (2007a) having reported BSE cases or having substantial risk associated with BSE include Albania, Austria, Belgium, Bosnia-Herzegovina, Bulgaria, Croatia, Czech Republic, Denmark, Federal Republic of Yugoslavia, Finland, France, Germany, Greece, Hungary, Ireland, Israel, Italy, Liechtenstein, Luxembourg, former Yugoslavia Republic of Macedonia, The Netherlands, Norway, Oman, Poland, Portugal, Romania, Slovak Republic, Slovenia, Spain, Sweden, Switzerland, Japan, and United Kingdom (Great Britain, including Northern Ireland and the Falkland Islands).

To facilitate importation of cattle and cattle products from countries, such as Canada, that presented a minimal risk of introduction of BSE into the United States the regulations were amended in 2005 to only exclude the importation of live bovines older than 30 months and products from slaughtered cattle older than 30 months (APHIS, 2005). The FDA has further expanded the restrictions by prohibiting the use of the following cattle material in human food and cosmetics:

- cattle material from nonambulatory, disabled cattle
- cattle material from organs from cattle 30 months of age or older in which infectious prions are most likely to occur, and the tonsils and the distal ileum of the small intestine of cattle of all ages
- cattle material from mechanically separated beef and
- cattle material from cattle that are not inspected and passed for human consumption

Moreover, documentation of compliance with these requirements has been mandated (FDA, 2006a).

Emergence of other food safety issues has likewise necessitated that the United States implement specific import restrictions to ensure the public's

Table 2 Summary of FDA imported food inspection activities

Year	No. of food entries under FDA's jurisdiction (millions)	No. of security reviews	Field exams n	Field exams %	Laboratory samples analyzed n	Laboratory samples analyzed %
1985[a]	0.9	–	–	–	–	–
1998[a]	3.0	–	–	–	–	–
2003[b]	6.0	–	78,659	1.3	25,736	0.4
2004[c]	7.5	33,111	70,926	0.9	24,480	0.3
2005[d]	8.7	86,187	84,997	1.0	25,549	0.3
2006[e]	8.9	89,034	94,545	1.1	20,662	0.2
2007[e]	9.1 (projected)	–	–	–	–	–

[a]Source: FDA (1998b).
[b]Source: FDA (2005).
[c]Source: FDA (2006b).
[d]Source: FDA (2007b).
[e]Source: FDA (2007c).

health. For example, an executive order signed by the President banned the importation of birds and unprocessed bird products from countries with the highly pathogenic avian influenza H5N1 virus (White House, 2005). Such import restrictions do not apply to processed products that have been rendered noninfectious.

Ensuring the safety of imported food is a daunting task because Americans consume a continually increasing amount of imported food (GAO, 1998b). The United States is moving from a nation self-sufficient in its food supply to one that is increasingly dependent on other countries (Gilmore, 2004). For example, imports in 2006 accounted for about 16% of the total vegetable supply and about 44% of the total U.S. fruit supply (see chapter 1). The quantity of imported food is escalating (Table 2), while the FDA's resources to inspect them are not keeping up (FDA Week, 2006; FDA, 2007d).

The United States' regulation of imported food has been criticized for many years, with the U.S. Government Accountability Office (GAO, formerly the General Accounting Office) a frequent critic (GAO, 1998a). This criticism crescendos as consumption of imported food rises (FDA, 2007d). A number of recent food-borne illness outbreaks illustrate that imported foods can challenge the U.S. regulatory system:

- In 1996, >1,465 cases of *Cyclospora* food-borne illness were reported in the United States and Canada, with Guatemalan raspberries identified as the most likely source (Calvin et al., 2002).

- In 1997, > 200 schoolchildren and teachers in Michigan contracted hepatitis A from strawberries shipped from Mexico (Hutin et al., 1999; Calvin, 2003).
- In 2000, 47 people became ill from *Salmonella* Poona, with cantaloupe from Mexico implicated as the source (Anderson et al., 2002).
- In 2002, 58 became ill from *Salmonella* Poona in the United States and Canada, with Mexican cantaloupe implicated (Anderson et al., 2002).
- In 2003, 555 people fell ill and three people died of hepatitis A associated with green onions imported from Mexico (CDC, 2003).

Currently, FDA inspects approximately one percent of the food imported under its jurisdiction. In 2003, Congress increased funding for import inspections, and in that year FDA hit a high of 1.3 percent of food imports inspected (*FDA Week*, 2006). Nonetheless, in 2004, the acting FDA Commissioner told Congress that the Agency's border inspectors were being swamped by the increasing number of imports and predicted that the Agency would inspect <1% of food imports in 2007 (*FDA Week*, 2006).

While import regulation by USDA relies primarily on its determination that a country has an equivalent food safety regulatory system, nonetheless, USDA checks every import shipment of meat, poultry, and egg products (at least for paper compliance). Therefore, concern arises over whether FDA's inspection resources are adequate. The FDA, however, lacks any statutory authority that would allow the agency to require exporting countries to have inspection systems equivalent to the United States. GAO has criticized this lack of authority (GAO, 1998a, 2008).

The FDA is criticized for the small percentage ($\approx$1%) of imported items inspected (Table 2) (Anonymous, 2007). Critics of the FDA system point out the much higher percentage (8 to 15%) of meat and poultry products that is reinspected (Table 3) by USDA field inspectors. However, the different systems employed by these agencies to track their work load (food entries versus pounds of product) makes comparison difficult, and the number of USDA food entries being inspected likely represents a substantially smaller number than the number of FDA inspections.

In addition, the FDA system of food safety regulation—for both domestic and imported food—relies on a small number of inspections. What is most important is not review of the numbers or percentages of inspections by FDA, which are small for both domestic and imported food, but rather, that domestically FDA has regulatory authority over the processing of food, not just the final product as with most imports. Thus, a spotlight on the small

Table 3 Summary of USDA imported food inspection activities[a]

Year	Meat and poultry under USDA's jurisdiction (million lbs)[a]	Reinspections[b]	
		million lbs	%[c]
2004	4,211	323	7.7
2005	4,303	414	9.6
2006	3,888	598	15.4
2007	3,949[d]	337	11.4

[a]Sources: FSIS (2004, 2005, 2006, 2007b).
[b]USDA designates these exams as reinspections because product was subjected to initial inspection in country of origin.
[c]Percent of total imported meat and poultry.
[d]First three quarters of FY 2007.

number of FDA import inspections draws attention away from the important point that port-of-entry inspections alone may never provide acceptable protection. A 1991 report by the Advisory Committee on the Food and Drug Administration called point-of-entry inspections an "anachronism" and considered the process of inspecting a final product to ensure conformity to standards "totally discredited" as a means of ensuring regulatory compliance (DHHS, 1991). It is widely accepted that a prevention-based system, such as Hazard Analysis and Critical Control Point (HACCP), is more effective and efficient at ensuring food safety. Simply put, prevention better ensures safety than end product testing alone.

A prevention-based inspection system would require inspection (verification) of hazard analysis and risk controls in the country of origin. Unfortunately, the FDA lacks the statutory authority to impose an equivalency requirement for importation of foods, generally has no review authority in or over foreign countries, and must rely on inspections at the U.S. ports of entry to determine the safety. Although the FDA applies a risk-based approach for targeting inspections of import shipments, the use of the term "risk based" is somewhat misleading because the FDA lacks the statutory authority to apply a scientific risk-based approach to imported food safety, because this would require authority to reach back to the country of origin.

CONCLUSION

While the law requires that all imported foods meet the same food safety standards as foods produced in the United States do, recent events have raised questions about the adequacy of the regulation of imported foods. The daunting task of ensuring the safety of imported food has grown as the United States moves from a nation self-sufficient in its food supply to one that increasingly depends on other countries.

The FSIS checks the documentation for each shipment of meat, poultry, and egg products imported to the United States. A portion of these shipments receives a more complete inspection. However, the FSIS primary means of ensuring the safety of imported meat, poultry, and egg products is through FSIS determination that the exporting country has an inspection system for the products that is equivalent to the U.S. system. This equivalency approach is more efficient and can be more effective than primary reliance on port-of-entry inspections.

The FDA lacks the statutory authority to impose an equivalency requirement for the importation of FDA-regulated foods. The FDA must rely on inspections at the U.S. ports of entry to determine the safety of the imported foods. Currently, the FDA checks <1% of the food imported under its jurisdiction. In addition, only a portion of the shipments checked receives a more complete inspection and laboratory tests. Consequently, concern arises that this inspection system is inadequate to assure imported food safety.

Finally, FSIS-regulated food shipments must be held at FSIS-registered warehouses for FSIS inspection until these shipments are released or refused entry. However, the FDA allows import shipments to be moved out of FDA control, which can result in unscrupulous importers evading FDA regulation by substituting products when inspection or sampling is called, or by simply selling the product before inspection.

REFERENCES

Anderson, J., S. Stenzel, K. Smith, B Labus, P. Rowley, S. Shoenfeld, L. Gaul, A. Ellis, M. Fyfe, H. Bangura, J. Varma, and J. Painter. 2002. Multistate outbreaks of *Salmonella* serotype *poona* infections associated with eating cantaloupe from Mexico—United States and Canada, 2000–2002. *Morb. Mortal. Wkly. Rep.* 51:1044–1047.

Anonymous. 2007. The dangers of imported food. *The Week* 7(315):15.

Calvin, L. 2003. Produce, food safety, and international trade: response to U.S. foodborne illness outbreaks associated with imported produce. *In* J. C. Buzby (ed.), *International Trade and Food Safety: Economic Theory and Case Studies*. U.S. Department of Agriculture, Economic Research Service, Agriculture Economic report no. 828 (AER828). [Online.] http://www.ers.usda.gov/Publications/AER828/. Accessed 14 January 2007.

Calvin, L., W. Foster, L. Solorzano, J. Mooney, L. Flores, and V. Barrios. 2002. Response to a food safety problem in produce: a case study of a cyclosporiasis outbreak. *In* B. Krissoff, M. Bohman, and J. Caswell (ed.), *Global Food Trade and Consumer Demand for Quality*. Kluwer Academic/Plenum Publishers, New York, NY.

Caribbean Produce Exchange, Inc. v. Secretary of Health and Human Services, 893 F.2d 3 (1st Cir. 1989).

Centers for Disease Control and Prevention (CDC). 2003. Hepatitis A outbreak associated with green onions at a restaurant—Monaca, Pennsylvania, 2003. *Morb. Mortal. Wkly. Rep.* 52:1155–1157.

Continental Seafoods, Inc. v. Schweiker, 674 F.2d 38 (D.C. Cir. 1982).

FDA Week. 2006. Vol. 12, no. 33 (Aug. 18). Inside Washington Publishers, Washington, DC.

Food and Agriculture Organization and the World Health Organization (FAO/WHO). 2002. BSE as a national and trans-boundary food safety emergency. FAO/WHO Global Forum of Food Safety Regulators. Marrakesh, Morocco, 28–30 January 2002. [Online.] http://www.fao.org/docrep/meeting/004/y2038e.htm. Accessed 24 September 2007.

Food and Agriculture Organization and the World Health Organization (FAO/WHO). 2005. Understanding the codex alimentarius. Revised and updated. [Online.] http://www.fao.org/docrep/008/y7867e/y7867e00.htm. Accessed 24 September 2007.

Fortin, N. *Food Regulation: Law, Science, Policy, and Practice*, in press. John Wiley & Sons, Inc., Hoboken, NJ.

General Accounting Office (GAO). 1998a. Food safety: federal efforts to ensure the safety of imported foods are inconsistent and unreliable. GAO/RCED-98-103, 30 April 1998, and GAO/T-RCED-98-191, 14 May 1998.

General Accounting Office (GAO). 1998b. Food safety: weak and inconsistently applied controls allow unsafe imported food to enter U.S. Commerce, Statement of Lawrence J. Dyckman, Director, Food and Agriculture Issues, Resources, Community, and Economic Development Division, Testimony before the Permanent Subcommittee on Investigations, Committee on Governmental Affairs, U.S. Senate. GAO/T-RCED-98-271, September 10, 1998.

General Accountability Office (GAO). 2005. Oversight of food safety activities. Federal agencies should pursue opportunities to reduce overlap and better leverage resources. GAO-05-213, March. [Online.] http://www.gao.gov/new.items/d05213.pdf. Accessed 31 January 2008.

General Accountability Office (GAO). 2007. David M. Walker, Comptroller General of the United States, Testimony before the Subcommittee on Agriculture, Rural Development, FDA, and Related Agencies, Committee on Appropriations, House of Representatives. Federal Oversight of Food Safety: High-Risk Designation Can Bring Needed Attention to Fragmented System. GAO-07-449T.

General Accountability Office (GAO). 2008. Federal oversight of food safety. FDA's Food Protection Plan proposes positive first steps, but capacity to carry them out is critical. GAO-08-435T, January 29. [Online.] http://www.gao.gov/new.items/d08435t.pdf. Accessed 31 January 2008.

Gilmore, R. 2004. US food safety under siege? *Nat. Biotech.* 22:1503–1505.

Goodwin v. United States, 371 F. Supp. 433 (SD Cal 1972).

Hutin, Y., V. Pool, E. Cramer, O. Nainan, J. Weth, I. Williams, S. Goldstein, K. Gensheimer, B. Bell, C. Shapiro, M. Alter, H. Margolis, and the National Hepatitis Investigation Team. 1999. A multistate, foodborne outbreak of hepatitis A. *N. Engl. J. Med.* 340:595–602.

Meserey v. United States, 447 F. Supp. 548 (D. Nev. 1977).

Pure Food and Drug Act of 1906, United States Statutes at Large (59th Cong., Sess. I, Ch. 3915, p. 768–772).

Seabrook Intl. Foods, Inc. v. Harris, 501 F. Supp. 1086 (DDC 1980).

U.S. Department of Agriculture, Animal and Plant Health Inspection Service (APHIS). 2005. Bovine spongiform encephalopathy; minimal-risk regions and importation of commodities: final rule and notice. Fed. Regist. **70**:9CFR Parts 93, 94, 95, and 96. [Online.] http://frwebgate.access.gpo.gov/cgi-bin/getdoc.cgi?dbname=2005_register&docid=fr04ja05-17. Accessed 24 September 2007.

U.S. Department of Agriculture, Food Safety Inspection Service (FSIS). 2004. Quarterly enforcement report. Table 3a. Imported meat and poultry products. Pounds of product presented, reinspected, and refused entry by fiscal year quarter (p. 11). [Online.] http://www.fsis.usda.gov/PDF/QER_Q4_FY2004.pdf. Accessed 17 September 2004.

U.S. Department of Agriculture, Food Safety Inspection Service (FSIS). 2005. Quarterly enforcement report. Table 3a. Imported meat and poultry products. Pounds of product presented, reinspected, and refused entry by fiscal year quarter (p. 12). [Online.] http://www.fsis.usda.gov/PDF/QER_Q4_FY2005.pdf. Accessed 17 September 2007.

U.S. Department of Agriculture, Food Safety Inspection Service (FSIS). 2006. Quarterly enforcement report. Table 3a. Imported meat and poultry products. Pounds of product presented, reinspected, and refused entry by fiscal year quarter (p. 12). [Online.] http://www.fsis.usda.gov/PDF/QER_Q4_FY2006.pdf. Accessed 17 September 2007.

U.S. Department of Agriculture, Food Safety Inspection Service (FSIS). 2007a. Eligible Foreign Establishments, updated June 5. [Online.] http://www.fsis.usda.gov/regulations_&_policies/Eligible_Foreign_Establishments/index.asp. Accessed 9June 2007.

U.S. Department of Agriculture, Food Safety Inspection Service FSIS). 2007b. Quarterly enforcement report. Table 3a. Imported meat and poultry products. Pounds of product presented, reinspected, and refused entry by fiscal year quarter (p. 12). [Online.] http://www.fsis.usda.gov/PDF/QER_Q3_FY2007.pdf. Accessed 17 September 2007.

U.S. Department of Commerce, National Oceanic and Atmospheric Administration (USDC, NOAA). 2007. USDC Seafood Inspection Program. [Online.] http://seafood.nmfs.noaa.gov/publications.htm Accessed 25 March 2007.

U.S. Department of Health and Human Services (DHHS). 1991. Final Report of the Advisory Committee on the Food and Drug Administration, Advisory Committee on the Food and Drug Administration (May).

U.S. Food and Drug Administration (FDA). 1979. FDA Memorandum of Understanding with Customs Service, Fed. Regist. **44**:53577 (14 September 1979).

U.S. Food and Drug Administration (FDA). 1996. FDA import procedures. Industry Activities Staff Flyer. [Online.] http://www.cfsan.fda.gov/~lrd/import.html. Accessed 7 February 2007.

U.S. Food and Drug Administration (FDA). 1998a. Guidance for industry: guide to minimize microbial food safety hazards for fresh fruits and vegetables. [Online.] http://www.foodsafety.gov/~dms/prodguid.html. Accessed 2 February 2007.

U.S. Food and Drug Administration (FDA). 1998b. Testimony statement of William B. Schultz before the United States Senate, 24 September 1998. [Online.] http://www.fda.gov/ola/1998/imported.htm. 25 Accessed September 2007.

U.S. Food and Drug Administration (FDA). 2002. Regulatory Procedures Manual, Chapter 9: Import Operations/Actions. [Online.] http://www.fda.gov/ora/compliance_ref/rpm_new2/ch9/default.htm. Accessed 24 September 2007.

U.S. Food and Drug Administration (FDA). 2005. Food and Drug Administration 2005 Budget. Foods, program activity data. [Online.] http://www.fda.gov/oc/oms/ofm/budget/2005/FOODS.htm. Accessed 25 September 2007.

U.S. Food and Drug Administration (FDA). 2006a. Recordkeeping requirements for human food and cosmetics manufactured from, processed with, or otherwise containing, material from cattle. Fed. Reg. 71(196): 21 CFR Parts 189 and 700. Available at http://www.cfsan.fda.gov/~lrd/fr061011.html. Accessed September 24, 2007.

U.S. Food and Drug Administration (FDA). 2006b. Office of Management. Budget Formulation and Presentation. 2006 Budget. [Online.] http://www.fda.gov/oc/oms/ofm/budget/2006/PDFs/Summary/Pages300thru335.pdf. Accessed 25 September 2007.

U.S. Food and Drug Administration FDA). 2007a. What countries have reported cases of BSE or are considered to have a substantial risk associated with BSE? [Online.] http://www.cfsan.fda.gov/~comm/bsefaq.html. Accessed 24 September 2007.

U.S. Food and Drug Administration (FDA). 2007b. Office of Management. Budget Formulation and Presentation. 2007 Budget. [Online.] http://www.fda.gov/oc/oms/ofm/budget/2007/HTML/1Foods.htm. Accessed 25 September 2007.

U.S. Food and Drug Administration (FDA). 2007c. CFSAN 2008 Budget Narrative. [Online.] http://www.fda.gov/oc/oms/ofm/budget/2008/1-BudgetNarrativeCFSAN.pdf. Accessed 25 September 2007.

U.S. Food and Drug Administration FDA). 2007d. FDA science and mission at risk. Report of the Subcommittee on Science and Technology. [Online.] http://www.fda.gov/ohrms/dockets/ac/07/briefing/2007-4329b_02_01_FDA%20Report%20on%20science%20and%20Technology.pdf. Accessed 31 January 2008.

White House. 2005. Executive order: Amendment to E.O. 13295 relating to certain influenza viruses and quarantinable communicable diseases. [Online.] http://www.whitehouse.gov/news/releases/2005/04/20050401-6.html Accessed 18 September 2007.

Imported Foods: Microbiological Issues and Challenges
Edited by Michael P. Doyle and Marilyn C. Erickson
© 2008 ASM Press, Washington, DC

Outbreaks of Food-Borne Diseases Related to the International Food Trade

Robert V. Tauxe, Sarah J. O'Brien, and Martyn Kirk

3

INTRODUCTION

Despite advances in food safety over many years, food-borne disease remains an important global public health problem. In the United States, contaminated food has been estimated to cause 76 million cases of illness each year, 323,000 hospitalizations, and 5,000 deaths (Mead et al., 1999). In England and Wales, food-borne disease causes an estimated 1.3 million cases, approximately 21,000 hospitalizations (accounting for some 88,500 bed days), and nearly 500 deaths annually (Adak et al., 2002), and in Australia, about 5.4 million cases, 15,000 hospitalizations, and 120 deaths (Hall et al., 2005; Australian Government Department of Health and Ageing [AGDHA], 2006). In many countries, a substantial proportion of foods are imported, and the volume of imported foods is increasing as patterns of production, trade, and taste change. While the greater part of the burden is no doubt from domestically produced foods, a fraction of the burden of food-borne disease is related to foods imported from other countries. Though data with which to estimate that fraction are sparse, the safety of imported foods is likely to be a growing challenge as the volume of imports increases. Detecting and investigating the outbreaks of food-borne illness related to imported foods and correcting the problems that lead to their contamination are likely to benefit both importing and exporting countries and should be an integral part of global food safety.

Robert V. Tauxe, Deputy Director, Division of Foodborne, Bacterial and Mycotic Diseases, National Center for Zoonotic, Vectorborne and Enteric Diseases, Centers for Disease Control and Prevention, 1600 Clifton Road, Mailstop C-09, Atlanta, GA 30333. Sarah J. O'Brien, Professor of Health Sciences and Epidemiology, University of Manchester, Clinical Sciences Building, Hope Hospital, Stott Lane, Salford M6 8HD, United Kingdom. Martyn Kirk, OzFoodNet, Office of Health Protection, Department of Health and Ageing, GPO Box 9848, MDP 14, Canberra 2601, Australian Capital Territory, Australia.

Imported foods are often widely distributed, so that an outbreak detected in one location may herald similar outbreaks in other locations or countries. For example, in 1983, a large outbreak of enterotoxigenic *Escherichia coli* infections was traced to raw-milk Brie cheese from France; similar outbreaks from the same cheese were then also reported in Denmark, Norway, and The Netherlands (MacDonald et al., 1985). In 1989, a local outbreak of staphylococcal food poisoning at a college cafeteria was linked to canned mushrooms from China and proved to be the first of four such outbreaks definitely linked to the same source and three others likely to have been associated (Levine et al., 1996). Subsequent actions in China in response to this investigation may have greatly improved canning safety, as no similar subsequent events have been reported. In 1994, a large outbreak of *Salmonella enterica* serotype Stanley infections linked to alfalfa sprouts in Finland provided the critical hypothesis that explained a similar outbreak in the United States, and both outbreaks were traced to alfalfa seeds from the same European distributor (Mahon et al., 1997). These outbreaks heralded the beginning of the large and ongoing problem of alfalfa sprout-associated infections that are often linked to seeds sold globally as agricultural commodities (Taormina et al., 1999).

A second international benefit of investigating these outbreaks is that it can draw the attention of food safety authorities in the exporting country to unsuspected food safety problems that may also be affecting their own population. For example, in 1991, a small outbreak of cholera in the state of Maryland was traced to a commercial frozen-fresh coconut milk product that was imported from Thailand (Taylor et al., 1993). Not only was the presence of *Vibrio* in that food a surprise to the Thai food safety authorities, but the manufacturer of the product was unknown to them and thus had escaped regulatory attention altogether. That same year, three months after the beginning of the Latin American cholera epidemic, two outbreaks of cholera in the United States were linked to frozen crabs brought informally from Ecuador in travelers' suitcases (Centers for Disease Control and Prevention [CDC], 1991b; Finelli et al., 1992). This was a preview of events that took place later that year when the investigation of epidemic cholera in Ecuador showed that undercooked crabs were a source of the infection there (Weber et al., 1994). In 1995, an international outbreak of *Salmonella* Agona infections in the United Kingdom and North America was linked to a savory peanut-based snack (Killalea et al., 1996). The snack was produced in Israel, where an ongoing outbreak of *Salmonella* Agona infections had not yet been explained; as a result of the international investigation, the outbreak in the exporting country was controlled (Shohat et al., 1996).

These outbreaks illustrate the speed with which new pathogens and disease syndromes can spread around the world, propelled by trade, global migration, and social change (Mead and Mintz, 1996). Detecting, investigating, and ultimately preventing these multinational outbreaks require close coordination among public health and food safety authorities across many jurisdictions (Tauxe and Hughes, 1996; Ammon and Tauxe, 2007). We describe the recent experiences in three countries with food-borne outbreaks traced to imported foods to highlight the challenges such investigations pose and the lessons they have to offer.

The Expanding Trade in International Foods

Long-distance transport of foodstuffs is probably as old as human civilization but has accelerated in recent times (Kiple, 2007). In the United States, the year-round pursuit of fresh and minimally processed foods offers the consumer an increasing array of foods imported from other countries. The produce aisles are full of items that have been produced and packed in warmer parts of the world, including many developing nations. In 1980, only 5 to 6% of U.S. fresh and frozen produce was imported. By 2001, 11% of the entire U.S. food supply by weight was imported, including 12% of beef, 17% of fresh and frozen vegetables, 23% of fresh and frozen fruit, and 83% of fresh and frozen seafood (Jerrardo, 2003). Mexico was the source of 27% of fruit imports, and other Latin American countries provided an additional 40%. One limitation on such imports is concern about fruit flies that would be harmful to orchards if they were introduced into the United States. This barrier can be overcome by fumigation, irradiation, or other treatments that kill fruit fly larvae present on the fruit. Many fresh fruits and vegetables are imported from countries that historically have been risky places to eat fresh produce (CDC, 2005). The importation of fresh meats is more restricted because of the risk that infectious animal diseases that are present in other countries, such as foot-and-mouth disease or African swine fever, might enter the country along with the meat and harm farm animal production in the United States. As a result, the fresh and frozen meat imported into the United States has come largely from Canada and Australia, which have well-established animal disease control programs, and recently from Argentina, after that country successfully eradicated foot-and-mouth disease.

In the United States, two separate federal agencies regulate the safety of foods. The Food Safety and Inspection Service of the U.S. Department of Agriculture (USDA) is responsible for meat, poultry, and pasteurized egg products, while the Food and Drug Administration (FDA) is responsible for all other foods. The two agencies have substantially different approaches to imported food safety due to their different enabling legislation (Anonymous, 2002). The USDA maintains agricultural attachés in other countries to evaluate the production of meat, poultry, and egg products destined for importation into the United States, requires certification that the safety inspection system under which the food was produced is equivalent to that of the United States, and inspects each shipment before it enters the United States. In contrast, the FDA does not require that the food be produced under a safety system equivalent to that of the United States, and instead relies on intermittent product testing at the border for chemical and biological contamination that would render the foods unsafe, inspecting and examining a small fraction of the imported foods under their jurisdiction.

The United Kingdom imports about $21.1 billion worth of food and live animals, of which 63% comes from the European Union (EU). Meats account for the largest percentage of these imports at $3.9 billion, with 78% of these coming in from the EU. In 2005, approximately 42% of the total food supply in the United Kingdom was not produced there, by value (Department of Environment, Food, and Rural Affairs [DEFRA], 2007). A number of government departments, agencies, local authorities, and inspectorates are involved in the regulation of legitimate trade and the control of illegal trade in animals, fish, plants, and their products into the United Kingdom. Overall, four government departments, DEFRA, the Foods Standards Agency, Her Majesty's Revenue and Customs, and the Forestry Commission, eight separate Inspectorates, and about 60 local authorities at ports are responsible for the control of legal and illegal trade in these products in England and Wales (Cabinet Office, 2002).

Much of the detailed legislation on food standards in the United Kingdom originates in the EU. Some products can only come into the EU through specific ports. Imported products of animal origin (POAO) must be presented at a designated border inspection point (BIP) for veterinary checks, and consignments failing the documentary, identity, or physical checks are not allowed into the United Kingdom and can be destroyed. Animal products entering the United Kingdom from other EU Member States must have undergone import checks at a BIP where they entered the EU. Examples of animal products imported for human consumption that must be checked include:

- meat, including fresh meat, meat products, minced meat, meat preparations, poultry meat, rabbit, farmed game meat, and wild game meat
- eggs and egg products
- fish and fishery products
- milk and milk products
- honey
- gelatin and gelatin products

Once a consignment has passed veterinary checks, a Common Veterinary Entry Document is issued and the POAO is permitted free circulation within the EU. This means that food products can be moved freely within the EU without customs checks, although there may be national controls where there are risks to public health. Consignments failing veterinary checks must be re-exported outside of the European Community or destroyed.

Most food not of animal origin may enter through any port, although importers are required to check that the port has the necessary facilities to handle food. Some products from specific countries are subject to emergency controls and can only enter into the United Kingdom through designated ports. These include products likely to be contaminated with aflatoxins (such as nuts) and uncultivated wild mushrooms that may be contaminated with radiocesium following the Chernobyl power station incident.

Australia imported AUD$125.9 million worth of food and live animals for food production in 2005 to 2006. In Australia, importation of food is regulated by the Australian Quarantine and Inspection Service (AQIS), which receives advice from Food Standards Australia New Zealand. The quarantine and inspection system is very robust and has led to Australia being largely free of zoonotic diseases important to trading status or animal production, such as foot-and-mouth disease and infectious bursal disease of chickens (Tanner, 1997). New Zealand has similarly robust systems of animal quarantine, which has prevented the introduction of *Salmonella* Typhimurium Definitive Type (DT) 104 into animal production systems (Crump et al., 2001). This has protected consumers from food-borne diseases. For example, Australian layer flocks are free of endemic *Salmonella enterica* serotype Enteritidis that can infect the internal contents of eggs and has resulted in significant epidemics in many other countries (OzFoodNet Working Group, 2006). In the past 20 years, there has been a large increase in the variety and volume of foodstuffs imported into Australia, particularly from neighboring Asian countries. This has led to outbreaks of food-borne disease associated with Asian food products, such as individually quick-frozen oysters and dried peanuts.

The Art and Practice of Food-Borne Outbreak Investigations

Food-borne outbreak investigations are an important public health activity at local, regional, and national levels. Most outbreaks are localized and typically are investigated by local authorities. Regional and national authorities assist in investigations of outbreaks that are particularly large, severe, unusual, or dispersed across several jurisdictions. These investigations are an integral part of the food safety system, because they provide an opportunity to learn much about the nature and source of the contamination, particularly when a specific food is implicated and traced back to its source. With the collaboration of food safety authorities, a detailed public health investigation can reveal how contamination might have occurred, and how it might be prevented. The goals of such an investigation are, first, to understand and control the immediate hazard and, second, to reconstruct how the outbreak happened well enough to ensure that similar events can be prevented in the future.

The nature of the outbreaks that are detected depends critically on the methods and vigor of ongoing public health surveillance. Unusual clusters of similar illnesses may first be noted as a result of reports from citizens, physicians, or clinical microbiologists. Such clusters are typically localized and may reflect a local contamination problem. Broader surveillance networks, which include further characterization and subtyping of strains of microbes isolated from ill persons, permit the detection of more dispersed outbreaks, even though they are less apparent in any one location. Such surveillance networks began with *Salmonella* serotyping, which is now a routine part of public health surveillance in more than 60 countries (Herikstad et al., 2002). This has been expanded to include phage typing and molecular subtyping by pulsed-field gel electrophoresis (PFGE) and other methods. The use of standard molecular subtyping in the PulseNet network for food-borne bacterial pathogens in the United States, the Enter-net network in Europe, and of other PulseNet networks around the world is greatly increasing the capacity to detect and to investigate dispersed outbreaks of bacterial food-borne pathogens (Swaminathan et al., 2006; Tauxe, 2006).

After surveillance identifies a cluster of possibly related illnesses, investigation of that cluster often requires labor-intensive epidemiologic efforts to identify the food that was the likely vehicle of transmission. These efforts typically include case-finding, hypothesis generation, systematic interview of ill persons and healthy controls, statistical comparison of the results of those interviews to identify the food or other exposures associated with illness, sampling of leftover foods of particular interest, and determination of exactly how and where the food was obtained and prepared. The successful investigation typically links the illness to exposure to a specific food or ingredient, with a statistical evaluation of the likelihood that such an association would be observed by chance alone. In the United States, a network of state and national epidemiologists, called OutbreakNet, coordinates multijurisdictional food-borne outbreak investigations. While it can be very helpful to identify the outbreak pathogen in the implicated food, it is not necessary to do so in order to take regulatory action. In the United States, a well-conducted epidemiologic investigation is a convincing basis for action. A strong epidemiologic association includes more than statistical probability. The investigation may also reveal that those who ate more of the implicated food are more likely to be ill, that the distribution of the implicated food and the illnesses are congruent, or that it is biologically plausible that the food as prepared could serve as the vehicle of infection.

Many investigations stop as soon as a particular food or food setting is implicated. However, when the same illness affected persons eating the food in many different parts of the country, then it is likely that the food must have been contaminated before it reached the final kitchen or store, and a traceback investigation may be conducted to trace the origin back though the food production chain, and if possible identify the likely point at which it became contaminated. The successful traceback investigation depends on the resources available, the quality of records kept along the distribution chain, the cooperation of the industry, and the nature of the food itself. Contamination may have occurred at any of several points from farm to table. Reconstructing how contamination occurred can require a complex, multidisciplinary assessment of the chain of production, processing, and distribution. When the contaminated food comes from another country, this phase of the outbreak investigation depends on the collaboration of the health and agricultural authorities in that country, even though there may be no recognized cases of illness in that country.

There are many hurdles to linking an outbreak firmly to an imported food. The investigation first needs to implicate a specific food item and clarify that contamination most likely occurred before the final kitchen. Then the origin of the contaminated food needs to be traced, which demands resources that are sometimes not available to the investigative team. The actual origin of a food is often obscured by the time it reaches the final kitchen, as domestic and imported products may be mixed together and relabeled by intermediary distributors. The report may neglect to include the origin even if it is known. For all these reasons, the small number of reported outbreaks is just the visible tip of a larger problem. They certainly underrepresent the frequency of outbreaks related to imported foods and should not be interpreted quantitatively as a direct measure of the risk of imported foods.

RECENT OUTBREAK SURVEILLANCE DATA

Many national public health authorities collect reports of investigated outbreaks of food-borne diseases from the local and regional health departments, and publish the results of this surveillance periodically. Collating summary reports of investigated outbreaks of food-borne disease is an important means of attributing illness to foods and developing sensible food policy (O'Brien et al., 2006). However, few reports include information about the nation of origin of the implicated food.

In the United States, approximately 1,200 food-borne outbreak investigations are reported annually to the CDC, currently through a web-based reporting system known as the Electronic Foodborne Outbreak Reporting

System (eFORS) (Lynch et al., 2006). Each report includes some basic information about the size of the outbreak, the etiology, whether microbial or chemical, the implicated foods, and the locations where the food was prepared and consumed, along with other optional additional information. An implicated food vehicle was reported for 62% of outbreaks between 1998 and 2002 (Lynch et al., 2006). Since 1998, a box on the report form can be ticked to indicate that the implicated food was imported, though this information is rarely reported. Between 1998 and 2004, 13 outbreaks were reported in this way to be caused by the importation of contaminated foods. To those, one may add six other investigations in which direct CDC participation occurred, and the implicated foods are known to be imported, for a total of 19 outbreaks between 1998 and 2004 (Table 1).

The United Kingdom participates in Enter-net, which was created in 1994 (Fisher, 1999). The network stemmed from a growing realization that, as barriers to trade across Europe came down and foreign travel became more popular and affordable, the potential for international outbreaks had increased. It was funded by the European Commission and, initially, its aim was to prevent human salmonellosis within the EU by strengthening international laboratory-based surveillance and by creating a regularly updated European *Salmonella* database open to all participants that could be used for monitoring outbreaks. It was formed as an adjunct to national surveillance schemes, not as a replacement for them. Proof of principle followed quickly, with the recognition and investigation of international outbreaks of *Salmonella enterica* serotypes Agona, Tosamanga, and Dublin (Hastings et al., 1996; Killalea et al., 1996; Shohat et al., 1996; Vaillant et al., 1996). Transatlantic collaboration also occurred during an outbreak of *Salmonella enterica* serotype Stanley infections, affecting the United States and Finland (Mahon et al., 1997). Timely exchange of information between experts in different EU countries led to effective public health action in Europe and beyond (Fisher, 1999). Table 2 illustrates the range of outbreaks identified and investigated through Enter-net between 1998 and 2002 that involved the United Kingdom and the country of origin of contaminated foods implicated in certain outbreaks. The country of origin of implicated foods is not routinely reported in outbreak surveillance in the United Kingdom.

In Australia, OzFoodNet is a surveillance network covering the entire country, which collects national data on all food-borne disease outbreaks (Ashbolt et al., 2002). OzFoodNet is modeled on the CDC FoodNet program of enhanced food-borne disease surveillance. Each year in Australia, there are approximately 100 outbreaks of food-borne disease that affect between 2,000 and 4,000 people. Between 1998 and 2004, there were eight outbreaks of food-borne disease associated with foods that were known to be

Table 1 Food-borne outbreaks in the United States and Australia traced to imported contaminated foods, 1998–2004[a]

Country	Year	No. of illnesses[b]	Pathogen	Implicated food(s)	Source country	References
United States	1998	9	Norovirus	Oysters	NR[c]	eFORS
	1999	79	*Salmonella* serotype Newport	Mango	Brazil	Sivapalasingam et al. (2003)
	1999	12	Unknown	Oysters	NR	eFORS
	1999	16	*Salmonella* serotype Typhi	Mamey fruit	Guatemala	eFORS
	1999	94	*Cyclospora*	Berries	Uncertain	eFORS (Herwaldt, 2000)
	2000	46	*Salmonella* serotype Poona	Cantaloupe	Mexico	CDC (2002a)
	2001	2	*Vibrio parahaemolyticus*	Crab salad	NR	eFORS
	2001	50	*Salmonella* serotype Poona	Cantaloupe	Mexico	CDC (2002a)
	2001	32	*Salmonella* serotype Kottbus	Alfalfa sprouts	NR	eFORS (Winthrop et al. (2003)
	2002	26	*Cyclospora*	Raspberries	Chile	eFORS
	2002	15	*Salmonella* serogroup C2	Fish soup	NR	eFORS
	2003	26	*Salmonella* serotype Poona	Cantaloupe	Mexico	CDC (2002a)
	2003	7	*E. coli* O157:NM	Alfalfa sprouts	Australia	eFORS
	2003	13	*E. coli* O157:NM	Alfalfa sprouts	Australia	eFORS
	2003	11	*Campylobacter*	Queso fresco cheese	Mexico	eFORS
	2003	945	Hepatitis A	Green onions	Mexico	eFORS (Amon et al., 2005)
	2003	12	*Listeria monocytogenes*	Queso fresco cheese	Mexico	eFORS
	2003	50	*Salmonella* serotype Typhimurium	Queso fresco cheese	Mexico	eFORS
	2003	12	*Salmonella* serotype Typhimurium	Queso fresco cheese	Mexico	eFORS
Australia	2001	23	*Salmonella* serotype Typhimurium DT104	Helva	Turkey	O'Grady et al. (2001)
	2001	55	*Salmonella* serotype Stanley	Dried peanuts	China	Kirk et al. (2004)
	2002	53	*Salmonella* serotype Montevideo	Tahini and helva	Egypt/Lebanon	Unicomb et al. (2005)

(continued)

Table 1 Food-borne outbreaks in the United States and Australia traced to imported contaminated foods, 1998–2004[a] *(continued)*

Country	Year	No. of illnesses[b]	Pathogen	Implicated food(s)	Source country	References
	2002	230	Suspected norovirus	Individually quick-frozen (IQF) oysters	Japan	OzFoodNet Working Group (2005)
	2003	5	*Salmonella* serotype Montevideo	Tahini and helva	Egypt/Lebanon	Unicomb et al. (2005)
	2003	17	Suspected norovirus	IQF oysters	Japan	OzFoodNet Working Group (2005)
	2003–2004	86	Norovirus	IQF oysters	Japan	Webby et al. (2007)
	2004	5	Suspected norovirus	IQF oysters	Japan	OzFoodNet Working Group (2005)

[a]The 1998 outbreak caused by *Shigella* and ETEC was inadvertently omitted from Table 1 and from subsequent summary statements about Table 1.
[b]Illness totals are for the reporting country or region, because the actual total number of cases globally is unknown. Four Hepatitis A outbreaks occurring within three months traced to green onions from the same Mexican locale are listed as one outbreak. Two outbreaks of *Salmonella* serotype Montevideo that were associated with consumption of the same brand of imported tahini and helva are listed as a single outbreak, as are three outbreaks of norovirus traced back to IQF oysters from a single harvest site in Japan.
[c]NR, not reported.

Table 2 International food-borne disease outbreaks investigated through Enter-net and involving the United Kingdom between 1998 and 2002[b]

Year	No.of illnesses	Pathogen	Countries involved[a]	Implicated vehicle (country of origin, if known)	Reference(s)
1998	>100	*Salmonella* serotype Newport	*England and Wales*, Finland	Cured ham	Lyytikainen et al. (2000)
2000	392	*Salmonella* serotype Typhimurium DT204b	England and Wales, Germany, *Iceland*, The Netherlands, Scotland	Lettuce	Crook et al. (2003)
2001	>100	*Salmonella* serotype Typhimurium DT104	Australia, Canada, England and Wales, Germany, Norway, *Sweden*	Helva (Turkey)	Andersson et al. (2001)
2001	>100	*Salmonella* serotype Stanley	*Australia, Canada*, England and Wales, Scotland	Peanuts (China)	Kirk et al. (2004)
2002	21	*E. coli* O157	*England and Wales*, France	Cucumber (Belgium)	Duffell et al. (2003)

[a]Countries that first reported an outbreak are in italics.
[b]Adapted from Fisher et al. (2005)

imported (Table 1). Some imported foods caused multiple point source outbreaks, which is a feature of these types of foods.

Tables 1 and 2 together list 30 distinct outbreaks caused by imported foods that were reported to or investigated by national authorities in the United States (19 outbreaks between 1998 and 2004), Australia (eight outbreaks between 1998 and 2004), and the United Kingdom (five outbreaks between 1998 and 2002). Note that two global outbreaks appear on both the United Kingdom and Australian lists and are only counted once: an outbreak of serotype Typhimurium DT104 infections traced to helva in 2001 and an outbreak of serotype Stanley infections traced to dried peanuts from China in 2001. Recurrent outbreaks suggest there is a persistent source or systemic problem in the exporting country, such as the series of *Salmonella enterica* serotype Poona outbreaks related to cantaloupes from Mexico, the recurrent sesame seed-related outbreaks from the Middle East, and the norovirus outbreaks related to raw shellfish from Japan.

The 30 outbreaks affected at least 2,634 persons. The number reported to be affected is a substantial underestimate, because often just the culture-confirmed cases are reported. For example, it has been estimated that there are 38 illnesses for every culture-confirmed case of salmonellosis that is reported (Voetsch et al., 2004). Furthermore, where the implicated product may have gone to many countries, the reported number of cases reflects those known to the reporting country. While these outbreaks surely underestimate the magnitude of the problem, they serve to illustrate the range of pathogens, foods, and sources involved. Among the 29 outbreaks for which a confirmed or suspected etiology was reported, 21 outbreaks were bacterial in origin (of which 14 were due to *Salmonella*), 6 were viral, and 2 were parasitic. The relative paucity of reported viral outbreaks should not be interpreted as a lower hazard, as national surveillance has greater sensitivity for detecting bacterial food-borne pathogens because of routine subtyping of *Salmonella* by serotyping, phage typing, and PFGE. Viral outbreaks accounted for 1,282 cases, 49% of the total, and bacterial outbreaks for 1,220 (46%), and parasitic outbreaks for 120 cases (5%).

The implicated foods were produce items in 17 outbreaks, seafood items in 8 outbreaks, dairy items in 4 outbreaks, and meat in a single outbreak. For the United States, all outbreaks were related to foods under the jurisdiction of the FDA. Produce was associated with 79% of all the outbreak-associated illnesses, seafood with 14%, meat with 4%, and dairy with 3%. For the 26 outbreaks in which the source of the contaminated food was determined, 12 came from Latin America, 8 from Asia/Pacific, 3 from the Middle East, and 1 each from Europe and Australia.

ILLUSTRATIVE OUTBREAKS

Published investigations provide illustrative details about the challenges of investigation and prevention.

Observations in the United States

Outbreaks from imported foods

Cyclosporiasis and Central American raspberries. In 1996, outbreaks of an unusual parasitic infection were identified in many parts of the United States and Canada. The causative organism, *Cyclospora cayetanensis*, had previously been identified as the cause of prolonged diarrheal illness in persons traveling to remote parts of the world, such as Katmandu, but was exceedingly rare in the United States. However, that year, 1,465 cases were reported from 20 states, the District of Columbia, and two Canadian provinces (Herwaldt et al., 1997). Many were part of 55 local outbreaks, whereas others were individual sporadic cases. Investigation of the local outbreaks quickly linked the illnesses to fresh raspberries from Guatemala, which had been hand picked and shipped by air in May and June of that year. Raspberries were not native to Guatemala and had recently been introduced as an export crop. Extensive investigations in Guatemala, in cooperation with the producer group and the Guatemalan authorities, identified possible routes of contamination on the berry farms, such as the use of untreated water to spray the berry plants with insecticides, but did not identify an animal reservoir or the definite mechanism of contamination for this pathogen. After local agricultural practices were modified, exports continued the following year, and similar outbreaks recurred. Further local investigation failed to identify clearly additional control measures that might be applied and so early in 1998, the importation of raspberries into the United States was halted. Imports to Canada continued, until later that same year Guatemalan raspberries again caused similar illness in Canada, and that country halted imports (Herwaldt, 2000). Additional fieldwork showed that a diarrheal illness that affected children in Guatemala each spring was also caused by *Cyclospora* and was associated with drinking untreated stream or river water (Bern et al., 1999). Whether that same water was also the source of contamination of the berries, or whether berries and water were both contaminated by some other source remains unclear. For now, one must conclude that this is the wrong place to grow raspberries. Curiously, blackberries are native to Central America and continue to be grown and exported from the same area without yet being linked definitively to a similar illness. In 1999, one outbreak was linked to mixed berries and, in one local outbreak that was part of this event, the risk was associated with raspberries from South America, suggesting that those fruits may be the source of infections in other parts of Latin America as well (Herwaldt, 2000). The *Cyclospora* problem illustrates how a virtually unknown disease can emerge when a new crop is introduced into a growing region, how swiftly a local and unrecognized problem can be shipped by air around a continent, and how, despite the best efforts of many to understand the ecology of contamination well enough to prevent it, it may be difficult to grow some foods safely in some places.

   Dysentery and parsley. In July and August 1998, two outbreaks of dysentery, a bloody and febrile diarrheal illness, affected persons at two separate social events in Minnesota. Both outbreaks were caused by *Shigella sonnei*. In one outbreak, standard epidemiologic investigations associated illness with eating foods made with parsley, and in the other, parsley was strongly suspected; both events shared the same parsley source (Naimi et al., 2003). The state health department used the PFGE method to fingerprint the bacterial DNA, as part of the newly formed national laboratory network, PulseNet, and showed that the *Shigella* strains from the two outbreaks had indistinguishable patterns. Because parsley is not often considered as a vehicle of food-borne infections and might be missed as a source of outbreaks, public health authorities around the continent were alerted. They identified six other outbreaks in three other states and two Canadian provinces caused by *S. sonnei* with the same PFGE fingerprint as the two outbreaks in Minnesota. In each of these outbreaks, once the hypothesis of parsley was suggested, the evidence led the investigators to either identify it as the source or to suspect it strongly. In addition, two other outbreaks of gastroenteritis in Minnesota were identified caused by a second microbe, toxigenic *E. coli*, better known as a cause of *turista* diarrhea among travelers to Mexico. These two outbreaks also were linked to parsley from the same distributor that supplied the parsley implicated in the shigellosis outbreaks. In all, 486 persons were ill with shigellosis and 77 with toxigenic *E. coli* diarrhea. The combination of two different enteric diseases from the same food item across multiple states and countries suggests a contamination scenario at or near the point of production that included a connection to raw sewage. The likely source of the parsley was traced back through distributors to a farm in Mexico. Investigators from the FDA, Mexico, and CDC visited that farm. Noteworthy observations included the use of water from the local municipal system to rinse and chill the parsley, and water that was likely to be unchlorinated to ice it. The chlorination of the local system was inadequate and intermittent, and the investigators were told that the farm workers themselves avoided drinking it, and specified that they be provided with bottled water to drink in their work contract. It is likely that fecal contamination of the local water supply had contaminated the parsley, creating a multinational outbreak of illness. Prevention measures for this scenario would be to guarantee the potability of the water used for washing and chilling the produce, for example, by installing an automated water chlorination device on site, with continuous monitoring and a failsafe switch that cuts off the water if the chlorine level drops. Though not required by regulatory statute everywhere in the United States, such equipment is used in many food processors and is an obvious part of good agricultural practice. It seems self-evident that water used to process fresh foods that are eaten without further cooking must be water that is safe to drink, though this is not a requirement in the United States.

Jaundice and green onions. In 2003, large outbreaks of hepatitis A infections affected customers of restaurants in Tennessee, Georgia, North Carolina, and Pennsylvania (Amon et al., 2005; Wheeler et al., 2005). In all, 1,023 cases were reported. In Pennsylvania, at least 124 of the 601 identified patients were hospitalized, and 3 died. The illness was typical, with jaundice, abdominal pain, fever, and prolonged malaise characteristic of this infection in adults. Approximately 15,000 persons received prophylactic gamma-globulin injections. By using DNA sequencing of the viral strains, it was found that the cases in the separate states were caused by three closely related strains of the virus that were very similar to those isolated from ill persons on the Texas-Mexico border. Like *Shigella*, food-borne hepatitis A infections are typically associated with an infected food preparer, but that seemed unlikely in these outbreaks, which affected many different restaurants, grocery stores, and nursing homes in the four states. Illness was linked to eating green onions, which in turn were traced back through the supply chain to four likely source farms in northern Mexico. In Mexico, hepatitis A is a common infection in young children, among whom it most often causes a relatively mild diarrheal illness. The initial investigation indicated that contamination was likely to have occurred at or soon after the time of harvest. On-site investigation with the FDA, Mexican authorities, and the CDC revealed issues of concern including a questionable quality of water used in packing sheds and for ice-making, poor sanitation and handwashing facilities, and the possibility that young children in diapers may have had contact with harvested produce. In this case, the preventive remedies may include ensuring that the water used for washing and icing the produce is disinfected and potable, and separating young children from the harvested food (FDA, 2003).

E. coli O157 infections and alfalfa seeds from Australia. Another outbreak illustrates that the source of an imported food problem is not always the developing world. Two related outbreaks of infection with *E. coli* O157:H- bacteria occurred in 2003 that were linked to consumption of alfalfa sprouts (Ferguson et al., 2005). The Minnesota outbreak of 7 cases occurred in January and February of that year, and the Colorado outbreak of 13 cases occurred in July and August; both were caused by strains of *E. coli* with indistinguishable PulseNet DNA fingerprint patterns, which had not previously been identified in the United States. For all but one case, traceback indicated a common source of seeds used by two sprouters. The seeds came from one seed producer in Australia. Alfalfa seeds are harvested in order to grow more alfalfa and are therefore regulated as agricultural commodities, not as food. The process of sprouting seeds greatly amplifies any bacterial enteric pathogens that are present (Taormina et al., 1999). Control efforts have depended on disinfecting seeds by dipping them in chlorine, and by conducting lot-by-lot testing of the water used to sprout the seeds. Apparently neither process was sufficient to protect the public in this case. After these outbreaks, the implicated seeds were diverted to agricultural uses. Although the source of the seeds was determined, the FDA reported that because there was no memorandum of understanding between Australia and the FDA that allowed the exchange of confidential commercial information, the traceback investigation did not extend to Australia (Ferguson et al., 2005). Unless specific conditions of contamination in the field can be identified and prevented, the preventive remedy for this food seems to be to apply disinfection strategies more thoroughly to the seeds and to expand more thorough lot-testing protocols at all sprouters.

Salmonella **serotype Poona infections and cantaloupes.** In the spring of 1991, a large multistate outbreak of *Salmonella* serotype Poona infections was linked to cantaloupes, the origin of which was not definitively established and may have been either Texas or northern Mexico (CDC, 1991a). More than 400 cases were confirmed in 23 states and 4 Canadian provinces. More recently, three successive multistate outbreaks occurred in the spring of 2000 (47 cases), 2001 (50 cases including 2 deaths), and 2002 (58 cases), each of which was linked to eating cantaloupes from farms in Mexico (CDC, 2002a). The PFGE patterns of the 2000 and 2002 outbreak strains were indistinguishable, whereas that of 2001 differed. The FDA evaluated the farms in 2000 and 2001, identifying several possible sources of contamination in fields and during washing and packing, finding that measures were insufficient to minimize microbial contamination, so that in 2001, melons from the producer were placed on a hold, effectively halting further imports. After the 2002 outbreak, the FDA issued an import alert on all cantaloupes from Mexico, effectively banning all imports, while the Mexican government developed a certification program under which farms would be permitted to resume exports. It was discovered subsequently that the cantaloupes in the 2002 outbreak were from the same producer but had been mislabeled to evade the 2001 producer-specific hold; the shipper pleaded guilty, and was fined $5 million. The recurrent outbreaks drove substantial regulatory and production level improvement in Mexico, and spring outbreaks of *Salmonella* serotype Poona infection have ceased. This indicates that the point of contamination has been controlled, while cantaloupe exports from Mexico continue.

Listeriosis and queso fresco. In 2003, the state of Texas reported to CDC 12 illnesses caused by *Listeria monocytogenes*, six of them confirmed by culture. These illnesses occurred in young pregnant women, causing severe illness in the fetus, and sometimes leading to miscarriage or stillbirth. The infections were linked to consuming a soft fresh cheese, *queso fresco*. Samples of cheese collected from informal vendors yielded *L. monocytogenes* on culture. The cheeses were made in Mexico from raw milk in homes and small firms, and brought over the border informally. In the past, when raw milk was diverted into informal cheese-making in the United States, this soft cheese had caused outbreaks of salmonellosis, listeriosis, and brucellosis (CDC, 1983; Villar et al., 1999; MacDonald et al., 2005). Recently, informal shipment of this cheese from Mexico was suspected as a source of pediatric bovine tuberculosis in New York City (CDC, 2006). The availability of low-cost overnight shipping services from small towns in Mexico means that cheese with "the taste of home" can be shipped rapidly anywhere in the United States. The remedy is long-term and multipronged and includes educating recent immigrants about the need for pasteurization of milk, developing greater control over the informal shipping of hazardous foods, and working with the Mexican and Central America authorities to bring the same messages to the rural population there.

Salmonella enterica **serotype Newport infections and mangoes: unintended consequence of a regulation.** One outbreak provides a cautionary tale about how phytosanitary food regulations can create a hazard. In 1999, a large multistate outbreak of infections caused by a strain of *Salmonella* serotype Newport affected at least 78 persons in 13 states (Sivapalasingam et al., 2003). Epidemiologic investigation, which focused on persons infected with the outbreak strain, defined by subtyping of the serotype in PulseNet, showed that illness was strongly associated with eating fresh mangoes. The mangoes were imported from Brazil, where they were produced at one large orchard that also exported mangoes to Europe. European colleagues were unable to find cases with the same strain. Investigation at the orchard and packing plant by the FDA, Brazilian authorities, and the CDC revealed that mangoes destined for export to the United States were treated routinely as required to prevent the introduction of the Mediterranean fruit fly. Mangoes destined for Europe were simply boxed, refrigerated, and shipped, without disinfestation, as the Mediterranean fruit fly is already native to Europe. The old standard fruit disinfestation treatment had been fumigation with ethylene dioxide gas. Out of concern for worker health and for the ozone layer, the USDA had in the 1990s required a shift to a new process, based on dipping the mangoes into a hot-water bath. This was followed by a cold-water bath to stop the heat from changing the taste of the mangoes. The hot-water bath was not chlorinated, and the cold-water bath was only chlorinated once a week, though it was open to the tropical environment and was in daily use. The sudden transfer of hot mangoes to cold water draws some of that water into the interior of the fruit as the interior of the mango cools and contracts, along with bacteria in the water (Penteado et al., 2004). The same phenomenon had already been well established for tomatoes, and is managed as a routine part of safety for that fruit in the United States (Rushing et al., 1996). However, the microbiologic implications of the sequential water dips had not been considered seriously for mangoes, and the required treatment also did not specify that water must be protected rigorously and disinfected at the point of use. One hopes that this remedy is now in place wherever the USDA mandates the disinfestation via a hot-water dip, that water, and any cooler water used thereafter, should be treated and monitored closely, so the process cannot itself become a source of contamination.

Outbreaks from exported foods

In addition, outbreaks have occurred in other countries that were traced to foods exported from the United States. Two distinct episodes are described below. These outbreaks illustrate how this country plays the roles of both source and recipient in the problem of global trade in contaminated foods.

E. coli O157 infections from ground beef in Japan. In February 2004, Japanese public health authorities identified three cases of *E. coli* O157:H7 infection in a family on Okinawa, who had eaten ground beef obtained from the U.S. military commissary there. Culture of the remaining frozen ground beef yielded *E. coli* O157:H7 with a newly identified PFGE pattern as determined by PulseNet Japan that matched the strains isolated from the family. Further case-finding efforts identified a fourth case in Okinawa, in a second family that purchased the same brand of ground beef from the same commissary. Notification of U.S. authorities led to the retrospective identification of two clinical isolates in California with the same PFGE pattern.

These infections may be associated, though the ground beef those patients consumed was not traced. The ground beef implicated in Japan had been produced in the United States six months earlier and distributed on the West Coast of the United States, as well as to military commissaries in the Far East. Identification of the cluster in Japan led to voluntary recall of approximately 90,000 pounds of beef from the military commissary system. This cluster highlights the utility of using the same subtyping protocols in different countries, as is now occurring through the PulseNet International (Swaminathan et al., 2006)

Salmonella enterica **serotype Enteritidis from almonds in Canada.** Between October 2000 and July 2001, 168 infections with an unusual phage type of *Salmonella* serotype Enteritidis were identified in North America, 157 in Canada and 11 in the United States. Investigation led by Canadian public health authorities implicated eating raw whole almonds (Isaacs et al., 2005). These almonds were sold and consumed without heating or blanching and were traced to specific almond groves in California, where the same unusual strain was identified in environmental samples. Almond trees are harvested by shaking them, so that the nuts fall to the ground, where they ripen for a week, and are then vacuumed up. Though most of the almond crop is heat treated before it is distributed, some consumers prize raw almonds for their taste and consistency. Further investigation indicated that *Salmonella* could easily grow in moist almond hulls, and that the nuts may easily be contaminated (Ueseqi and Harris, 2006). Almonds subsequently harvested from the implicated orchards were reportedly diverted to uses where they would be heat treated (CDC, 2004). In early 2004, a cluster of infections with *Salmonella* serotype Enteritidis, defined by PulseNet, led to the detection of a second outbreak with a different phage type, with 29 matching human isolates from 12 states and Canada. After this too was linked to consumption of raw almonds from a different California source, 13 million pounds of almonds were recalled by the producer (CDC, 2004). The same almonds were shipped to at least seven other countries. In 2003, almonds were California's largest agricultural crop and represented 80% of the global commercial almond production; 5% was sold for consumption raw. These recurrent international outbreaks, and the subsequent efforts to understand them, have led to plans that beginning with the 2007 harvest, all California almonds (including those for export) will either be treated to achieve a 4 log *Salmonella* reduction or will be labeled clearly as unpasteurized and in need of further treatment before consumption (Almond Board of California, 2006). Thus, the Canadian outbreak investigation combined with the pressure of a recurrent similar outbreak has stimulated a systematic improvement in the global food supply.

Observations in the United Kingdom
Outbreaks from imported foods
One of the lessons from the outbreaks presented below is that most were small when detected. This highlights the benefits of a comprehensive, centralized system for typing salmonellae in the United Kingdom, thereby increasing both the sensitivity and the specificity of the surveillance system.

Salmonella enterica **serotype Napoli in chocolate from Italy.** In 1982, 245 cases of *Salmonella* serotype Napoli infection occurred in an outbreak in England and Wales (Gill et al., 1983). Most case patients were children under 15 years of age who lived in the south of England. Fifty-one people were admitted to the hospital. Food histories obtained during hypothesis-generating interviews implicated two types of small chocolate bars. There was a strong association between consuming these two types of chocolate bars in a case-control study conducted among case and control households. The implicated chocolate-covered bars were imported from Italy, and both types were found to be contaminated with the outbreak strain. Four-fifths (32 tons) of a 3-million bar consignment of contaminated chocolates was immediately withdrawn from the market by the producer. Since 245 case patients had resulted from the sale of 600,000 bars, it is likely that approximately 200 hospital admissions from the many thousands of potential infections were prevented. Because the high-fat content of chocolate protects any salmonellae present from stomach acid, it is difficult to tolerate any detectable level of salmonellae in chocolate. This is especially important given that small children, who are particularly susceptible to food-borne illness, are avid chocolate eaters.

Shigella sonnei **in iceberg lettuce from Spain.** In June 1994, several small outbreaks of *S. sonnei* were detected in England and Wales, following an initial alert from Sweden where two outbreaks involving 52 people had occurred the previous month (Frost et al., 1995). Investigations in Sweden had implicated iceberg lettuce and peeled frozen prawns as vehicles of infection. Hypothesis-generating interviews conducted in England and Wales also suggested that iceberg lettuce was a vehicle of infection. This hypothesis was tested in a case-control study involving 28 case patients and 49 matched controls. The estimated odds ratio for consuming iceberg lettuce was 13.8. In sporadic cases associated with consumption of lettuce from particular commercial catering premises, the dates of onset were compared with the dates of delivery of iceberg lettuce by wholesalers. The distribution chain was traced back through importers supplying wholesale markets in England. These, in turn, were supplied by packers in Spain. These findings were consistent with those of investigators in Norway tackling a similar outbreak

(Kapperud et al., 1995). Although microbiologic investigations of iceberg lettuce during the second week of June 1994 did not yield *S. sonnei*, the iceberg lettuce season in Spain, which had begun in October, ended early in June. Thus, the source of lettuce available for testing at the time was not traceable. The extent of the outbreak was sizeable—between 14 June and 31 July 1994, 495 isolates of *S. sonnei* were referred to the national Laboratory of Enteric Pathogens for confirmation and further typing. Similarly to the experience in the United States, a considerable number of outbreaks linked to produce have occurred subsequently in the United Kingdom, but tracing contaminated salad, fruit, or vegetables has proved to be very difficult on a number of occasions. This difficulty in traceback means that identifying and controlling the contamination at the source has been problematic.

Salmonella enterica **serotype Agona in savory snack from Israel.** Between the beginning of December 1994 and the end of January 1995, 24 cases of infection with *Salmonella* serotype Agona phage type 15 were identified in England and Wales (Killalea et al., 1996). Most of the case patients were children and most had Jewish surnames. Hypothesis-generating interviews with the parents of eight primary household cases revealed that four of the children had eaten a kosher, peanut-flavored, ready-to-eat savory snack. The snack was made of maize and had been imported from Israel. The hypothesis was tested in a case-control study involving 15 case patients and 32 controls. There was an 88-fold increased risk of becoming ill after eating the implicated kosher snack. Microbiologic examination of packets of the savory snack obtained from retailers found *Salmonella* serotype Agona phage type 15. The contents of eight packets, subjected to quantitative studies, revealed that the estimated *Salmonella* count ranged from 2 to 45 CFU per 25-g packet. The outbreak also affected people in Israel, the United States, and Canada. In the United Kingdom, a Food Hazard Warning was issued and the product was withdrawn. This outbreak demonstrated dramatically the worldwide ramifications of an imported food contamination incident.

Salmonella enterica **serotype Anatum in infant formula from France.** Twenty-two cases of infection with *Salmonella* serotype Anatum were recognized in infants in England, Scotland, France, and Belgium between September 1996 and January 1997 (Anonymous, 1997). There were no deaths. The ages of the patients suggested strongly that a baby food was a possible vehicle for infection. In a case-control study, in which the mothers of the first 12 case infants (ages 2 to 8 months, mean age 4.9 months) and 40 control infants (ages 1 to 9 months, mean age 4.6 months) were interviewed by telephone, the odds ratio of illness following consumption of a particular brand of infant formula was 62. Molecular analysis of isolates from case infants revealed that the outbreak strain was characterized by the presence of a single plasmid of about 50 megadaltons (9 of 12 cases in the United Kingdom; 2 of 3 cases in France). The infant formula milk implicated in the outbreak was produced in a factory in France using dried milk that might have been sourced from either of two spray drying plants, one in France and the other in The Netherlands. The factory produced a range of baby foods for export to different countries, though the implicated formula milk was specially manufactured for the United Kingdom market. On 24 January 1997 the implicated product was withdrawn from sale in the United Kingdom. A fortnight later a batch of infant formula milk, which had been produced from the same raw materials as the United Kingdom product, was withdrawn from distribution in France. As a result of a comprehensive, centralized surveillance system, this outbreak was detected and controlled while it was still small.

Salmonella enterica **serotype Paratyphi B biovar Java in desiccated coconut from Malaysia.** In February 1999, the national Laboratory for Enteric Pathogens detected an increase in cases of *Salmonella enterica* serotype Paratyphi B biovar Java phage type Dundee. Infections were serious. Seven cases had been admitted to hospital and three had developed septicemia. Biovar Java is rare in England and Wales (on average, 15 cases per annum) so that so many cases in such a short time was highly unusual. Biovar Java is indistinguishable from serotype Paratyphi B serologically and is identified on the basis of biochemical tests. Though biovar Java usually causes typical nontyphoidal salmonellosis, in this outbreak the clinical presentation in the patients affected was compatible with the more severe enteric fever caused by serotype Paratyphi B. Case findings revealed 17 cases across the south of England. There was a preponderance of Asian surnames among those affected, and half of the cases, who had been admitted to hospitals, had significant underlying medical conditions. In an unmatched case-control study of 16 cases and 31 controls, the risk of illness was associated with shopping at a particular outlet (outlet A) in north London. No particular food vehicle was, however, identified. A breakthrough occurred when colleagues in Austria reported that they had isolated the pathogen from desiccated coconut imported from Malaysia. Food and environmental swabs from outlet A were examined at the London Food, Water and Environment Laboratory, and two samples of desiccated coconut yielded biovar Java phage type Dundee. Using PFGE, the isolates were indistinguishable from those from the case patients. Further investigations at the warehouse that supplied outlet A revealed that a consignment of coconut from Malaysia was contaminated not only with biovar Java PT Dundee but also with biovar Java PT Battersea, and *Salmonella enterica* serotypes Mbandaka, Bareilly,

Senftenberg, Arizona, Weltevreden, and Hvittingfoss. The mean probable number of salmonellae per gram was 10. In total, 33 people were affected in this outbreak and nearly 11,000 kg of contaminated coconut were withdrawn from sale voluntarily. The circumstances that led to the contamination of the coconut were not fully understood, although a plausible hypothesis was that the consignment had been damaged at some point by flood water.

The unusual clinical presentation, the fact that an important proportion of cases had underlying medical conditions, and the policy to type all salmonellae at a central point led to swift identification of this outbreak. It was disappointing that the epidemiologic investigation failed to identify a food vehicle precisely, but not entirely surprising given the nature of the contaminated ingredient. However, the epidemiologic study pinpointed the location for the detailed food and environmental microbiologic investigations that yielded the source. This case illustrates strongly the power of combined epidemiologic, microbiologic, and environmental approaches to investigations.

Salmonella enterica **serotype Enteritidis in eggs from Spain.** A national outbreak of serotype Enteritidis infection, affecting a total of 530 laboratory-confirmed cases, occurred between June 2003 and February 2004 (I. A. Gillespie, unpublished report). Approximately 10% of case patients reported foreign travel before the onset of symptoms. In a case-control study of primary sporadic United Kingdom-acquired case patients, consumption of eggs outside the home (OR = 5.02; 95%CI 2.09 to 12.05; $P > 0.001$) was independently associated with illness.

Seven local clusters were identified and investigated as part of the national outbreak. The eggs implicated in at least four clusters were supplied, either directly or indirectly, by one egg-packing station/supplier that was sourcing eggs from the United Kingdom and the EU, mainly Spain. *Salmonella* serotype Enteritidis PT14b isolated from patients in all four outbreaks, and from eggs from Spain sampled in two outbreaks, were indistinguishable by plasmid profiling and PFGE.

Between 1 January 2003 and 19 November 2004, the Health Protection Agency (HPA) received 94 reports of local outbreaks of *Salmonella* serotype Enteritidis infection. In 91 outbreaks for which typing data were available, almost three-quarters were due to *Salmonella* serotype Enteritidis of phage types other than PT4, which had previously predominated in the United Kingdom: PT14b (19 outbreaks; 28%) and PT1 (10 [7 NxCpL]; 15%), PT6 (5; 7%) and PT8 (5; 7%) were the most common types. Most outbreaks of *Salmonella* serotype Enteritidis infection in 2003 and 2004 were linked to catering premises and restaurants serving Chinese food (43/75; 57%), which was the single most common type of premises reported. In 23 of 53 outbreaks that were thought to be egg-related outbreaks, the eggs were known (Brown et al., 2001) or suspected (Anonymous, 2006) by the investigators to have originated in Spain. United Kingdom eggs were implicated in 13 outbreaks; eggs from a number of countries were implicated in five outbreaks, and in 14 outbreaks the source country of the eggs was unknown. In outbreaks where eggs from Spain were implicated, PT14b, PT59, PT1, and PT6d were most often reported.

Between October 2002, when outbreaks were first recognized, and November 2004, 16,971 eggs were sampled and *Salmonella* spp. were recovered from 3.4%. *Salmonella* was isolated from 5.5% of Spanish eggs and 6.3% of eggs of unknown origin used in catering premises linked to outbreaks. This was significantly higher than the level of contamination found in non-Lion Quality United Kingdom eggs sampled (1.1%) (Little et al., 2007).

The United Kingdom produces approximately 880 million dozen hen eggs annually. Fifteen million dozen are sold abroad, mainly to the EU. Between 1980 and 1999, non-United Kingdom eggs as a proportion of all eggs available in the United Kingdom, excluding those sold abroad, increased from 4% to 9%. From 2000 to 2003, eggs sourced from abroad increased from 13% to 17% of United Kingdom production. By 2003 most non-United Kingdom eggs were sourced from Spain (33%), with Germany (22%) and The Netherlands (20%) also important sources (HPA, 2004b).

In all, the HPA investigated over 80 outbreaks, with at least 2,000 confirmed cases. The Food Standards Agency liaised with its Spanish counterparts to initiate investigations and controls in Spain. An action plan to improve the quality of eggs was drawn up by the Spanish authorities (HPA, 2004a).

Outbreak from exported foods

There seem to be few examples where contaminated food exported from the United Kingdom has caused outbreaks. This apparent anomaly might simply reflect the variations in the way that surveillance is conducted across the EU, with which the United Kingdom trades the most. If the countries to which the United Kingdom exports have less sensitive surveillance systems, then it is possible that outbreaks that might have been linked to contaminated food from the United Kingdom go unrecognized.

Variant Creutzfeldt-Jakob disease and British beef. A controversial example of a food contamination problem exported from the United Kingdom is bovine spongiform encephalopathy (BSE). BSE was identified in the British cattle population in 1986 (Wilesmith et al., 1992). The epidemic affected nearly 200,000 cattle, and its legacy is an outbreak of human variant Creutzfeldt-Jakob disease (vCJD), thought to have resulted from consuming beef products contaminated by central nervous system tissue (Brown et al.,

2001). Since 1994 when vCJD was recognized, there have been about 10 to 15 cases per year, but the future size and geographic distribution of cases in countries that have imported infected British cattle or cattle products, or have endogenous BSE, is still unpredictable.

France has the second highest number of vCJD cases worldwide (Chadeau-Hyam and Alperovitch, 2005). Imports of bovine carcasses from the United Kingdom probably constituted the main source of exposure of the French population to the BSE agent, and meat products consumed while visiting the United Kingdom were also a possible source of exposure. In a modeling exercise, the number of future vCJD cases in France was estimated using a simulation approach. Both the distribution of the vCJD incubation period and the age-dependent susceptibility to the BSE agent were estimated from United Kingdom data. Thirty-three future cases of vCJD were estimated, suggesting that a large vCJD epidemic in France is very unlikely. Since France (to which 60% of the total British exports of bovine carcasses were sent) was highly exposed to the BSE agent, those results should reassure countries worldwide.

A worldwide ban on the export of British beef was introduced by the EU in 1996 in response to a statement by the British Government that BSE was a possible cause of ten cases of vCJD (McKee and Steyger, 1997). This focused attention on the power of the EU to restrict trade on grounds of public health. For trade in goods to be limited on grounds of public health, several criteria must be met. First, there must be real grounds for concern that a threat to health exists. Second, the action must be proportionate to the objective being pursued and should not be achievable in another way not requiring a restriction on trade, such as labeling. Third, the restriction must be a "seriously considered health policy" and must be necessary to protect health while going no further than it is necessary to do so. Finally, if a ban is imposed, it is incumbent on the state imposing it to monitor the situation and rescind it if it becomes clear that the threat to health no longer exists (McKee and Steyger, 1997).

A central tenet of the Treaties that form the basis of European law is to proscribe any attempt by a Member State to act in any way that inhibits free movement of goods (one of the so-called "four freedoms" of movement—of goods, capital, persons, and services) (McKee and Steyger, 1997). As the EU enlarges, the four freedoms of movement are likely to have even wider ramifications for the control of food-borne disease.

Observations in Australia

Outbreaks from imported goods

Salmonella enterica serotype Typhimurium DT 104 associated with imported helva. In 2001, one state, Victoria, reported a cluster of 23 cases of infection with multi-drug-resistant *Salmonella* serotype Typhimurium DT

104, which is rare in Australia. Initial hypothesis-generating interviews revealed that cases were predominantly of Turkish ethnicity and had consumed food from delicatessens. A source was not identified for the cluster of cases, despite a case-control study exploring the association with various foods purchased from delicatessens. Swedish epidemiologists published a report of an outbreak of *Salmonella* serotype Typhimurium DT 104 concerning an association with helva, a confectionary made from sesame seed paste, which had originated in Turkey (Andersson et al., 2001). After seeing this report, personnel from the Victoria department of health reinterviewed cases and controls and found a strong association with the two brands of helva, which on culture yielded various serotypes of *Salmonella* (O'Grady et al., 2001). The original source of contamination in Turkey was never identified despite inquiries with the Turkish ministry of health. The full extent of this outbreak was not documented, but initial reports following the outbreak suggested it affected several European countries.

Salmonella enterica **serotype Stanley associated with imported dried peanuts.** In 2001, health departments in several Australian states noted an increase in cases of *Salmonella* serotype Stanley infection that was fully sensitive to antibiotics. In Australia, infections with *Salmonella* serotype Stanley are normally associated with travel to Asia and are often drug resistant. Hypothesis-generating interviews revealed that cases were predominantly people of Asian ethnicity who had not traveled. One or two patients reported eating, prior to illness, a dried peanut product originating in China. Based on initial interviews, a laboratory tested the dried peanut product and isolated serotypes Stanley and Newport. Further interviews of cases revealed that >50% of case patients had consumed these peanuts. A rapid report was disseminated on Promed, which assisted both the United Kingdom and Canada to identify related cases and contaminated product (Kirk, 2001). The outbreak prompted a recall of product on three continents, although the source of contamination was never identified in the exporting country. The international investigation team shared PFGE images electronically instead of sending isolates for comparison to a single laboratory (Kirk et al., 2004).

Salmonella enterica **serotype Montevideo in sesame seed products.** Sesame seed products caused additional outbreaks of salmonellosis in Australia during 2002 and 2003 (Unicomb et al., 2005). The first outbreak of *Salmonella* serotype Montevideo infections occurred in one Australian state, New South Wales, and affected 53 persons. Patients had commonly consumed tahini originating from Egypt, which on culture subsequently yielded *S. enterica* serotype Montevideo and other serotypes. Testing also identified other batches of tahini and helva that yielded *Salmonella* spp. As a result of the outbreaks, tahini was recalled and an international alert posted in Eurosurveillance (Unicomb et al., 2003). Following the international alert, Canada and the United Kingdom identified contaminated tahini at the retail level, which precipitated product recalls. No source was identified for contamination in the exporting country. A second outbreak was identified in Australia in June to July 2003 and another in New Zealand in August 2003. These smaller outbreaks of *Salmonella* serotype Montevideo infections were due to tahini originating from Lebanon. The multinational outbreak investigation team recommended a review of the controls for *Salmonella* spp. during the production of sesame-based products and an upgrading of the risk status for foods containing sesame seeds.

Norovirus outbreaks linked to individually quick-frozen oysters. Six outbreaks of gastroenteritis occurred between the years 2002 and 2004 following consumption of oysters imported from Japan. Three of these were confirmed outbreaks of norovirus infection, while the remainder was suspected to be caused by that agent. The oyster products were sold individually quick frozen (IQF), a process in which oyster meat is shucked before rapid freezing and sold in batches of 1 kg or more. In some instances, the outbreaks occurred where the food service industry served these oysters raw in cocktails, despite packaging that recommended to "cook before consumption." In late 2003 and early 2004, three outbreaks occurred because of three different brands of IQF oysters in three different Australian States, including one where the oysters were cooked before consumption (OzFoodNet Working Group, 2005). A nationally coordinated investigation traced these products to the same company and harvest area in Japan. Multiple strains of norovirus were detected in the stools of infected patients and oyster products (Webby et al., 2007).

The oyster products were withdrawn from the Australian marketplace, but an additional outbreak several months later due to the same batch prompted the Australian Quarantine Inspection Service to restrict imports of IQF oysters harvested in certain growing areas in Japan and Korea. The restriction of imports from Korean growing areas was based on outbreaks reported from New Zealand some years earlier. Despite New Zealand experiencing outbreaks with Korean oysters they continued to allow importation and experienced a large outbreak in 2006 associated with Korean oysters consumed at an international rugby test match (G. Simmons, personal communication). This outbreak prompted New Zealand to review conditions of importation for these products. These types of outbreaks are probably common in Korea and Japan, because 153 (54%) of 287 of food-borne norovirus outbreaks occurring in Japan during a 34-month surveillance period from 2000 to 2003 were associated with oysters (Anonymous, 2003).

Outbreaks from exported foods

Outbreaks associated with sprout seed. Internationally, sprout seeds have caused several outbreaks of *Salmonella* and enterohemorrhagic *E. coli* infections. The majority of outbreaks have been caused by contaminated alfalfa sprouts, and in earlier years, mung beans. Australia is a large supplier of the world market for sprout seeds, with most produced to grow more alfalfa for feeding livestock. Though these seeds are not a food in themselves, they become one once they are sprouted. Australian sprout seed has been implicated in outbreaks of pathogens where the pathogen is rare in the importing country, but common in Australia. These include nonmotile strains of *E. coli* O157 and *Salmonella enterica* serotype Kottbus (CDC, 2002b; Winthrop et al., 2003; Ferguson et al., 2005). Australian authorities have been reluctant to investigate local seed-growing practices without evidence of Australian outbreaks. In 2005, Australia experienced two outbreaks of *Salmonella enterica* serotype Oranienberg associated with locally produced sprout seeds. These outbreaks raised concerns about the safety of seeds produced largely to feed animals and resulted in considerable work with industry to improve safety associated with these products (OzFoodNet Working Group, 2006).

CURRENT EPIDEMIOLOGIC CHALLENGES

1. Detecting outbreaks associated with internationally distributed foods can be difficult when cases are widely dispersed. The capacity in any one jurisdiction to detect such outbreaks depends on its overall public health capacity for surveillance and response to food-borne infections. As with any food-borne outbreak, detection will depend on a combination of laboratory subtyping of strains, alert epidemiologists and microbiologists at local, regional, and national levels, and the ability to rapidly combine information across jurisdictions. Detection can occur in several ways. A local outbreak may herald a wider event, as occurred in the episode of shigellosis from parsley. A rare disease may suddenly appear in clusters of many areas, as occurred with cyclosporiasis from raspberries. However, the most common way for these dispersed outbreaks to be detected is through pathogen subtyping in public health laboratories. Routine and standardized microbiologic subtyping with rapid comparison of the results across jurisdictions, as in the PulseNet and Salm-Gene networks, is the only way that many dispersed food-borne outbreaks can be detected (Tauxe, 2006).

2. Investigating the outbreaks can be complex for several reasons. One of the greatest enemies of the epidemiologist is the time that elapses

between onset of illness in the cases and the recognition of an outbreak. In retrospective studies of illness, recall bias among case patients may limit information on food exposures. The longer the time that elapses between illness onset and investigation, the greater the chance that individuals may simply forget exactly what they ate. Thus, retrospective studies might, in fact, capture food preferences, rather than the actual food exposure. Similarly, time delays are also the enemy of the microbiologist. Leftover food will have been discarded and products with short shelf lives will have disappeared and not be available for testing.

Even if case patients can remember what they ate, they may be unaware of the contaminated constituents that made up the implicated meal. Epidemiologic investigations identify food vehicles, but the source of contamination might be an ingredient of the food that might not be obvious to the consumer. For example, a commodity like desiccated coconut can be an ingredient of lots of food items. In certain circumstances, like the decoration on a cake, this will be obvious. In others, like its use in a curry sauce, it may be less so. The food vehicle, curry, might well be identified as a food vehicle in an outbreak. Similarly, coconut-covered cakes might be identified as food vehicles. But it takes considerable painstaking work to track down individual ingredients that make up those food items, and recognize a common link between cakes and curry. By the time an investigation is initiated, a contaminated batch might have been used up so that the ingredients available for microbiologic examination are not the same as those that were used in the implicated food. Where the food vehicle and/or organism are unusual, e.g., chocolate contaminated with *Salmonella* serotype Napoli, the investigation is often much more straightforward than an outbreak involving commonly used ingredients like shell eggs or chicken, and common serotypes like *Salmonella* serotype Enteritidis.

Finally, the sheer complexity of coordinating investigations across multiple jurisdictions can be challenging. The relatively small number of cases in any one place means that local resources may be difficult to mobilize, and the variations in investigative approaches, legal authorities, and even languages can complicate using a consistent approach to hypothesis generation, case interviews, and control selection. Just as standardized laboratory subtyping methods have greatly improved outbreak detection, so will investigative networks that develop and promote standard methods. Networks like OutbreakNet, OzFoodNet, and Enter-net are critical to successful investigations of geographically dispersed outbreaks.

3. Conducting traceback and investigation in the importing country can be difficult. Traceback investigations may be hampered by poor labeling of food, or no labeling at all, particularly on commodities like fresh produce. Food items may have passed through many intermediaries before appearing on shelves in shops or supermarkets. In an outbreak of infection with multiresistant *Salmonella* serotype Typhimurium DT 104 in the United Kingdom in which the implicated food vehicle was lettuce, it was possible to trace back the salad vegetable supply chain from case clusters, via retailers, caterers, wholesalers, and other middlemen to two wholesale markets in the West Midlands and one wholesale supplier in northwest England (Horby et al., 2003). Unfortunately, the complexity of the food supply chain and the lack of identifying markers on salad stuffs made it extremely difficult to track salad vegetables back to the original grower. In some cases, the supply chain was traced through five stages before reaching a firm that imported the salad items from a wholesale market on the European mainland. These long supply chains caused problems in tracing food, and labeling of fresh salad produce was not sufficient to allow proper tracing of products.

   Traceback may be further complicated because of product substitution and relabeling. During the outbreaks of *Salmonella* serotype Montevideo associated with tahini and helva in 2003 that affected Australia and New Zealand, there was evidence of relabeling product for sale in the marketplace (Unicomb et al., 2005). This can make tracing the product very difficult, including identifying the source in the originating country.

   Even if traceback is successful there are often delicate negotiations with the food safety authorities in the importing country. The level of proof implicating an imported food may be challenged by the source country, and some producers can be very uncooperative. Without such cooperation, it is difficult to learn how to prevent such contamination from happening again, and in the event of recurrence, the importing country may have little recourse than to halt future importations to protect the health of their public.

4. Collaboration with partners in other countries is critical. Managing international food-borne outbreaks relies on robust investigation in countries where disease occurs, and in the country where the implicated food is manufactured or grown. Just as in multijurisdictional investigations in one country, there are often difficulties in conducting international investigations due to variability in epidemiologic and laboratory methods, questionnaires, and priorities (Kirk et al., 2004). In some instances, the team in one country may have no epidemiologist involved, whereas in

others there may be no laboratory scientists. Leadership of the investigations may be unclear, particularly where multiple countries have cases.

The international molecular laboratory and investigative networks PulseNet and Salm-Gene have greatly improved investigations associated with internationally distributed foods, allowing better detection of dispersed outbreaks and rapid communication concerning outbreak strains in humans and foods. However, in many parts of the world there is no surveillance, and in others, surveillance of enteric pathogens is not systematic or of good quality (Swaminathan et al., 2006). The international adoption of the standard subtyping methods, including *Salmonella* serotyping, as promoted by WHO Global Salm Surv program, and the molecular subtyping methods of PulseNet will enhance collaborative detection and investigation efforts around the globe. Similarly, promoting standard epidemiologic methods for surveillance and investigation will increase the likelihood that national food safety authorities will respond to epidemiologic investigations conducted in other countries. The 30 national epidemiology training programs that are members of TEPHINET (Training Programs in Epidemiology and Public Health Interventions Network) and the 26 countries that participate in EPIET (the European Program for Intervention Epidemiology Training) are implementing and promoting such methods around the world (Table 3) (Moren et al., 1996; Anonymous, 2006).

Traditionally, disease investigators identifying imported foods through outbreak investigation have communicated to other countries about possible spread of the outbreak through public domain information sources. These include rapid publications, such as *Morbidity & Mortality Weekly Report*, *Eurosurveillance*, and public domain listservs, such as Promed mail. Another important means of identifying related cases in other countries is informal bilateral contact with neighboring countries, such as exists between Canada, the United States, and Mexico, or through formal arrangements, such as the Europe-wide network—Enter-net (Fisher and Threlfall, 2005). In recent years, the World Health Organization (WHO) has strengthened its involvement in emergency food safety issues through INFOSAN—the international food safety authorities network. During large outbreaks of food-borne disease associated with internationally distributed foods, INFOSAN has disseminated information to affected countries and assisted with inquiries to the source of contamination in the country producing the food. The revised 2005 International Health Regulations, which took effect on 15 June 2007, bring broader reporting requirements than

Table 3 Current members of the international epidemiologic training networks Training Programs in Epidemiology and Public Health Interventions Network (TEPHINET) and European Program for Intervention Epidemiology Training (EPIET)

TEPHINET	EPIET
Argentina	Austria
Australia	Belgium
Brazil	Cyprus
Canada	Czech Republic
Colombia	Denmark
Germany	Estonia
Ghana	Finland
Guatemala	France
Honduras	Germany
India	Greece
Italy	Hungary
Japan	Ireland
Kazakhstan	Italy
Kenya	Latvia
Korea	Lithuania
Kyrgyzstan	Malta
Malaysia	Norway
Mexico	Poland
Nicaragua	Portugal
Peru	Slovak Republic
Philippines	Slovenia
Saudi Arabia	Spain
Spain	Sweden
Tajikistan	Switzerland
Thailand	Netherlands
Turkmenistan	United Kingdom
Uganda	
United States	
Uzbekistan	
Zimbabwe	

Source: Moren et al., 1996; Anonymous, 2006.

before, including reporting any "public health emergency of international concern" (WHO, 2005; Baker and Fidler, 2006). This revised requirement means that large outbreaks of serious food-borne illness with important international ramifications are to be reported by member countries to WHO for dissemination through INFOSAN or similar systems. However, INFOSAN is largely concerned with distribution of contaminated food, and in many countries, INFOSAN typically only reaches the regulatory food safety officials, not the epidemiologists. WHO does not have capacity to coordinate investigations of clusters or suspected outbreaks of food-borne illness.

Implementing control measures in the country that is the source of the food can be difficult. Usually, inquiries are conducted as part of bilateral discussions between the country identifying the disease outbreak and the country supplying the food. It is rare for an ongoing source of contamination to be identified because of the time elapsed between the outbreak and subsequent traceback and food safety inquiries. Speedy and transparent investigation and successful control in the exporting country can bolster confidence in their products. Recurrent outbreaks are evidence of lack of control and can depress the export market.

5. Food may be brought in by informal transport. The noncommercial and gray market transport of foods across international borders can pose a challenge. In 1991, two outbreaks of cholera in the United States were caused by "suitcase seafood" brought informally from Ecuador (CDC, 1991b; Finelli et al., 1992; Weber et al., 1994). Though this is not regulated, it was discouraged through education efforts in the Hispanic press and the cooperation of the agricultural inspectors at the airports. Currently, the informal transport of queso fresco into the United States from Mexico is a public health hazard. This fresh soft cheese can be contaminated with a variety of pathogens when it is made from unpasteurized milk. The same cheese made from unpasteurized milk in the United States is sometimes marketed door to door or through other unregulated means, also causing outbreaks. Educating informal cheese producers in this country to first pasteurize the milk has had some local success (Villar et al., 1999). Informal importation of unsafe cheese seems likely to continue until milk is universally pasteurized in Mexico and Central America.

6. Trade in feedstuffs can play a role. Ingredients for animal feed are also traded internationally, and contamination of animal feed has led to dissemination of pathogens to the animals that receive the feed. This provides a mechanism by which even locally produced foods can be affected by imported pathogens (Crump et al., 2002). In 1969 and 1970, *Salmonella* serotype Agona emerged dramatically as a public health problem on three continents, becoming the second most common serotype in the United Kingdom and the eighth most common in the United States (Clark et al., 1973). Investigation of an outbreak at a restaurant in

Arkansas led back to a farm that provided the implicated poultry, and ultimately to imported Peruvian fishmeal that was an ingredient of the chicken feed. That contaminated fishmeal was the likely explanation for the global dissemination of *Salmonella* serotype Agona. As described above, exported animal protein used in animal feeds from the United Kingdom may have led to cases of both animal and human spongiform encephalopathy in other countries (Taylor and Woodgate, 1997). In 2007, an outbreak of kidney failure in pets in the United States was related to toxic wheat gluten imported from China, the nitrogen content of which had been boosted with melamine and other chemicals (FDA, 2007). Though the hazards of imported animal feed components lie outside the scope of this chapter, they underline the international links that connect all countries and that drive the global need for safe food.

TRANSLATING FINDINGS INTO PREVENTION CAN HELP EVERYONE

These illustrative outbreaks show the insights into prevention that can be developed when food safety investigations cross borders and the likely sources and circumstances of contamination are identified. In the long run, the insights from these investigations will guide and propel sustained prevention efforts around the world. An important part of the food safety system in each country is the public health capacity to conduct surveillance, to detect outbreaks, to investigate them in order to determine the source of contamination, and to develop and implement long-term prevention strategies. This capacity is critical to the improvement of all food safety in that country, including that of food exports. Encouraging and promoting more robust public health systems in exporting countries that can address the recurrent challenges of local and export-related foods will contribute to the safety of food globally.

As our food supply becomes more globalized, rapid and direct collaboration in surveillance, outbreak investigation, traceback, and prevention is critical (Tauxe and Hughes, 1996). International collaboration is the *sine qua non* of consequential investigation. In many of these outbreak investigations, close collaboration with the exporting country authorities, and with the industry itself, led to better prevention strategies for the long term. If a country can export food or feedstuffs to another country, then in the event of disease associated with that export, a way must be found to communicate directly with the food safety, public health, and agricultural authorities in that country, in order to investigate and resolve it.

Outbreaks of food-borne illness traced to an imported food can herald the emergence of a previously unusual pathogen or a contamination of an unsuspected product, which makes them particularly useful for bringing emerging food-borne challenges to light. Thus, the *Cyclospora* outbreaks described above focused new attention on what had previously been a poorly understood pathogen limited to the developing world. Identification of *Salmonella* serotype Enteritidis in exported raw almonds from the United States presaged a later outbreak in the United States with the same subtype from that previously unrecognized source. Because the first export-related outbreak had stimulated research into the ecology of and potential solutions to the problem of *Salmonella* in almonds, a technical solution was already under development that when implemented will reduce the risk of salmonellosis from almonds in general. The identification of alfalfa sprouts in Finland in association with *Salmonella* serotype Stanley infection drew new attention in several countries to the sprout problem, and *E. coli* O157 infections in the United States that were traced to imported alfalfa seeds from Australia were the beginning point for discussing further investigations, in advance of recognized sprout-associated outbreaks in that country. It is likely that the recurrent outbreaks of sesame seed-associated salmonellosis from the Middle East and of oyster-associated norovirus infections from Japan reflect ongoing health hazards in those areas and should serve to stimulate further control efforts that ultimately will benefit all.

Not surprisingly, the largest outbreaks are related to imported fresh fruits and vegetables that are eaten without further cooking. They illustrate how close attention to the general principles of food safety is critical everywhere, including the need for careful pasteurization, safe canning, good agricultural practices, good manufacturing practices, and the safety engineering approach of hazard analysis/critical control point (HACCP). In particular, the quality of the water used for rinsing, chilling, and other produce food treatment is critical and must be ensured with failsafe attention. Requirements for safety of exported foods should be no less than those required in the United States, and may indeed need to be stronger. Routine certification and verification programs that reach back to the production sites are needed to make sure that they are applied. This may take the form of purchase contract requirements imposed by the firms receiving the imports and regulations imposed by the food safety authorities in the exporting countries.

Placing an embargo on specific foods or specific producers is a tool of last resort, but as illustrated by the recurrent outbreaks of salmonellosis from cantaloupes and cyclosporiasis from raspberries, it can be an important strategy. It should be kept in mind that simply halting importation of an unsafe food into one country may cause the product to be diverted to other countries, so that the nationality of the victims is changed, while their number is not diminished. The terms under which such embargoes have been lifted include process certification and verification, which illustrates the more general regulatory principles that may be needed. Australia restricted imports of IQF oysters from certain growing areas in Japan and Korea based on the level of risk posed by these products. Australia restricted Korean oysters based on New Zealand's experience of outbreaks with these products. New Zealand continued to experience large outbreaks of norovirus due to these products following Australia's import restrictions.

Improving Disease Prevention in General

Food may be produced for export in countries where general levels of drinking water treatment, sanitation, and personal health are lower than in the industrialized world. This means that a higher level of concern is warranted about the health of the workers and their families. While it is difficult to say just where the *Shigella* bacteria on the parsley or the hepatitis A virus on the green onions came from, they must ultimately have been from feces of infected humans. Efforts to reduce infections among workers, their families, and the surrounding populations would be expected to have a general preventive effect. In fact, important changes in water and sewage sanitation are occurring in many parts of the developing world. For example, until 1991, raw municipal sewage was used in the city of Santiago, Chile, to irrigate the fields where the fresh produce was grown. In that year, spurred by a cholera epidemic, a sewage treatment plant was constructed. The following year, typhoid fever declined by 86% in Santiago, hepatitis by 54%, and cholera did not occur at all (Alcayaga et al., 1993). Provision of safe drinking water is both a basic need and a profound public health advance, provided either through well-maintained municipal systems, point-of-use disinfection strategies, or both (Mintz et al., 2001; Clasen and Mintz, 2004). Spurred by the Latin American cholera epidemic, Mexico embarked on a nationwide effort to reduce fecal transmission of *Vibrio* and other pathogens (Sepulveda et al., 2006a, b). The percentage of Mexican municipalities with potable water was 55% in 1990, rising to 85% in 1991 and 94% in 1994 (Sepulveda et al., 2006b). Education about handwashing and sanitation and improved access to medical care and preventive medicine also yielded huge benefits, particularly for children. As the program advanced, rates of pediatric diarrhea plummeted from 7.2 per 1,000 live births in 1990 to 1.1 per 1,000 live births in 2005, and overall pediatric death rates fell by half in the same period. The next "International Decade for Action: Water for Life" began in 2005, which marks the beginning of further efforts to promote access to safe drinking water around the world as part of the Millennium Development Goals for 2015 (Moe and Rheingans, 2006). National health improvement policies in other exporting countries similar to those undertaken in Chile and Mexico would be expected to make the foods we import from them safer as well.

As the *Cyclospora* saga shows, introducing new crops into subtropical and tropical areas may reveal new hazards not previously recognized in the exporting country. The potential for introduction of pathogens from local wildlife and insects, as well as from the local water supply, must be considered. As the mango-associated outbreak illustrates, regulations to control one part of a food safety problem that are developed without regard to microbiological circumstances can lead to contamination and illness themselves.

SUMMARY

Food-borne infections from imported foods will be a continuing problem, driven by the desires of consumers in the industrialized world for access to fresh produce year round, for exotic foods from elsewhere, and in immigrant communities, by their desire for the taste of home. Collaborative international approaches to detecting and investigating food-borne outbreaks are vital to controlling and preventing them at the source and are greatly aided by standardized laboratory subtyping methods for food-borne pathogens and by compatible epidemiologic approaches. Entirely new and unanticipated food safety threats can arise. Meeting the challenge of existing and emerging food-borne diseases depends on more robust public health systems in each country and good international collaboration in surveillance and investigation. One logical starting point for prevention is to require food safety processes in exporting nations to be at least the equivalent to those required for foods produced in the importing nation. However, all food safety systems are incomplete, and the risk of potential contamination is likely to be greater in the developing world because of general sanitary conditions. General strategies to improve the health of the workers and rural populations in those countries and to increase the capacity of their public health and food safety systems are likely to have long-term benefits to the health in those countries, as well as to prevent infections in the countries to which they export.

ACKNOWLEDGMENT

We thank Rachel Woodruff, MPH, for her assistance in reviewing the outbreak records from eFORS.

REFERENCES

Adak, G., S. Long, and S. O'Brien. 2002. Trends in indigenous foodborne disease and deaths, England and Wales: 1992 to 2000. *Gut* **51**:832–841.

Alcayaga, S., J. Alcayaga, and P. Gassibe. 1993. Cambios del perfil de morbilidad en algunas patologias de transmission enterica con posterioridad a un brote de colera. Servicio de Salud Metropolitano Sur. Chile. *Rev. Chil. Infectiol.* 1:5–10.

Almond Board of California. 2006. *Almond Almanac 2006.* Almond Board of California. [Online.] http://www.almondboard.com/files/PDFs/ALM%5F6060%5FAlmanac07lr.pdf. Accessed 29 April 2007.

Ammon, A., and R. V. Tauxe. 2007. Investigation of multi-national foodborne outbreaks in Europe: some challenges remain. *Epidemiol. Infect.* 135:887–889.

Amon, J., R. Devasia, G. Xia, O. Nainan, S. Hall, B. Lawson, J. Wolthuis, P. Macdonald, C. Shepard, I. Williams, G. Armstrong, J. Gabel, P. Erwin, L. Sheeler, W. Kuhnert, P. Patel, G. Vaughan, A. Weltman, A. Craig, B. Bell, and A. Fiore. 2005. Molecular epidemiology of foodborne hepatitis A outbreaks in the United States, 2003. *J. Infect. Dis.* 192:1323–1330.

Andersson, Y., B. de Jong, L. Hellström, U. Stamer, R. Wollin, and J. Giesecke. 2001. *Salmonella* Typhimurium outbreak in Sweden from contaminated jars of helva (or halva). *Eurosurveill. Wkly.* 3 [Online.] http://www.eurosurveillance.org/ew/2001/010719.asp. Accessed 4 January 2007.

Anonymous. 1997. Investigation Internationale: Belgique France Royaume-Uni Et Le Reseau Salm-Net / International Investigation: Belgium United Kingdom France and the Salm-Net Network. Preliminary report of an international outbreak of *Salmonella anatum* infection linked to an infant formula milk. *Eurosurveillance* 2:22–24.

Anonymous. 2002. Food import/export control and certification import systems of the United States of America. Second FAO/WHO Global Forum of Food Safety Regulators, Bangkok, Thailand, 12–14 October 2004. www.fao.org/docrep/meeting/008/ae162e.htm. Accessed 2 June 2007.

Anonymous. 2003. Outbreaks of norovirus infection, January 2000–October 2003. *Infect. Agents Surveill. Rep. (Japan)* 24:309–310.

Anonymous. 2006. Preface: global epidemiology. Proceedings of the third TEPHINET Conference, Beijing, China, November 8–12, 2004. *Morb. Mortal. Wkly. Rep.* 55(Suppl.): 1–2. (See also www.tephinet.org for current individual member roster; accessed October 24, 2007.)

Ashbolt, R., R. Givney, J. Gregory, G. Hall, R. Hundy, M. Kirk, I. McKay, L. Meuleners, G. Millard, J. Raupach, P. Roche, N. Prasopa-Plaizier, M. Sama, R. Stafford, N. Tomaska, L. Unicomb, C. Williams, and O. W. Group. 2002. Enhancing foodborne disease surveillance across Australia in 2001: The OzFoodNet Working Group. *Commun. Dis. Intell.* 26:375–406.

Australian Government Department of Health and Ageing (AGDHA). 2006. *The Annual Cost of Foodborne Illness in Australia.* AGDHA, Canberra.

Baker, M., and D. Fidler. 2006. Global public health surveillance under new international health regulations. *Emerg. Infect. Dis.* 12:1058–1065.

Bern, C., B. Hernandez, M. Lopez, M. Arrowwood, M. de Mejia, A. de Merida, A. Hightower, L. Venczel, B. Herwaldt, and R. Klein. 1999. Epidemiologic studies of *Cyclospora cayetanensis* in Guatemala. *Emerg. Infect. Dis.* 5:766–774.

Brown, P., R. Will, R. Bradley, D. Asher, and L. Detwiler-Brown. 2001. Bovine Spongiform Encephalopathy and variant Creutzfeldt-Jakob Disease: Background, evolution, and current concerns. *Emerg. Infect. Dis.* 7:6–16.

Cabinet Office. 2002. The organisation of the Government's controls of imports of animals, fish, plants and their products. [Online.] http://www.cabinetoffice.gov.uk/publications/reports/pdf/illegal%20imports%20paper.pdf. Accessed 24 July 2007.

Centers for Disease Control and Prevention (CDC). 1983. Epidemiologic notes and reports. Brucellosis—Texas. *Morb. Mortal. Wkly. Rep.* **32**:548–553.

Centers for Disease Control and Prevention (CDC). 1991a. Epidemiologic notes and reports. Multistate outbreak of *Salmonella poona* infections—United States and Canada, 1991. *Morb. Mortal. Wkly. Rep.* **40**:549–552.

Centers for Disease Control and Prevention (CDC). 1991b. Epidemiologic notes and reports. Cholera—New York 1991. *Morb. Mortal. Wkly. Rep.* **40**:516–518.

Centers for Disease Control and Prevention (CDC). 2002a. Multistate outbreaks of *Salmonella* serotype Poona infections associated with eating cantaloupe from Mexico—United States and Canada, 2000–2002. *Morb. Mortal. Wkly. Rep.* **51**:1044–1147.

Centers for Disease Control and Prevention (CDC). 2002b. Outbreak of *Salmonella* serotype Kottbus infections associated with eating alfalfa sprouts—Arizona, California, Colorado, and New Mexico, February–April 2001. *Morb. Mortal. Wkly. Rep.* **51**:7–9.

Centers for Disease Control and Prevention (CDC). 2004. Outbreak of *Salmonella* serotype Enteritidis infections associated with raw almonds—United States and Canada, 2003–2004. *Morb. Mortal. Wkly. Rep.* **52**:484–487.

Centers for Disease Control and Prevention (CDC). 2005. *Health Information for International Travel 2005–2006.* Elsevier, New York.

Centers for Disease Control and Prevention (CDC). 2006. Human tuberculosis caused by *Mycobacterium bovis*—New York City, 2001–2004. *Morb. Mortal. Wkly. Rep.* **54**:605–608.

Chadeau-Hyam, M., and A. Alperovitch. 2005. Risk of variant Creutzfeldt-Jakob disease in France. *Int. J. Epidemiol.* **34**:46–52.

Clark, G., A. Kauffman, E. Gangarosa, and M. Thompson. 1973. Epidemiology of an international outbreak of *Salmonella agona. Lancet* **2**:490–493.

Clasen, T. F., and E. D. Mintz. 2004. International network to promote household water treatment and safe storage. *Emerg. Infect. Dis.* **10**:1179–1180.

Crook, P., J. Aguilera, E. Threlfall, S. O'Brien, G. Sigmundsdottir, D. Wilson, I. Fisher, A. Ammon, H. Briem, J. Cowden, M. Locking, H. Tschaepe, W. Van Pelt, L. Ward, and M. Widdowson. 2003. A European outbreak of *Salmonella enterica* serotype Typhimurium definitive phage type 204b in 2000. *Clin. Microbiol. Infect.* **9**:839–845.

Crump, J., P. Griffin, and F. Angulo. 2002. Bacterial contamination of animal feed and its relationship to human foodborne illness. *Clin. Infect. Dis.* **35**:859–865.

Crump, J. A., D. R. Murdoch, and M. G. Baker. 2001. Emerging infectious diseases in an island ecosystem: The New Zealand perspective. *Emerg. Infect. Dis.* **7**:767–772.

Department of Environment, Food, and Rural Affairs (DEFRA). 2007. *Food Statistics Pocketbook.* [Online.] http://statistics.defra.gov.uk/esg/publications/pocketstats/foodpocketstats/fsiyp.pdf. Accessed 9 November 2007.

Duffell, E., E. Espie, T. Nichols, G. Adak, H. De Valk, K. Anderson, and J. Stuart. 2003. Investigation of an outbreak of *E. coli* O157 infections associated with a trip to France of schoolchildren from Somerset, England. *Eurosurveillance* **8**:81–86.

Food and Drug Administration (FDA). 2003. FDA update on recent hepatitis A outbreaks associated with green onions from Mexico. [Online.] http://www.fda.gov/bbs/topics/NEWS/2003/NEW00993.html. Accessed 4 February 2007.

Food and Drug Administration (FDA). 2007. Detention without physical examination of all vegetable protein products from China for animal or human food use due to the presence of melamine and/or melamine analogs. Import Alert no. 99-29, 27 April 2007 [Online.] http://www.fda.gov/ora/fiars/ora_import_ia9929.html. Accessed 3 June 2007.

Ferguson, D., J. Scheftel, A. Cronquist, K. Smith, A. Woo-Ming, E. Anderson, J. Knutsen, A. De, and K. Gershman. 2005. Temporally distinct *Escherichia coli* O157 outbreaks associated with alfalfa sprouts linked to a common seed source—Colorado and Minnesota, 2003. *Epidemiol. Infect.* **133:**439–447.

Finelli, L., D. Swerdlow, K. Mertz, H. Ragazzoni, and K. Spitalny. 1992. Outbreak of cholera associated with crab brought from an area with epidemic disease. *J. Infect. Dis.* **166:**1433–1435.

Fisher, I. 1999. The Enter-net international surveillance network—how it works. *Eurosurveillance* **4:**52–55.

Fisher, I., and E. Threlfall. 2005. The Enter-net and Salm-gene databases of foodborne bacterial pathogens that cause human infections in Europe and beyond: an international collaboration in surveillance and the development of intervention strategies. *Epidemiol. Infect.* **133:**1–7.

Frost, J., M. McEvoy, C. Bentley, Y. Andersson, and B. Rowe. 1995. An outbreak of *Shigella sonnei* Infection associated with consumption of iceberg lettuce. *Emerg. Infect. Dis.* **1:**26–29.

Gill, O., P. Sockett, C. Bartlett, M. Vaile, B. Rowe, R. Gilbert, C. Dulake, H. Murrell, and S. Salmaso. 1983. Outbreak of *Salmonella napoli* infection caused by contaminated chocolate bars. *Lancet* **321:**574–577.

Hall, G., M. Kirk, N. Becker, J. Gregory, L. Unicomb, G. Millard, R. Stafford, K. Lalor, and O. W. Group. 2005. Estimating foodborne gastroenteritis, Australia. *Emerg. Infect. Dis.* **11:**1257–1264.

Hastings, L., A. Burnens, B. de Jong, L. Ward, I. Fisher, J. Stuart, C. Bartlett, and B. Rowe. 1996. Salm-Net facilitates collaborative investigation of an outbreak of *Salmonella tosamanga* infection in Europe. *Commun. Dis. Rep.* **6:**R100–R102.

Herikstad, H., Y. Motarjemi, and R. V. Tauxe. 2002. Salmonella surveillance: a global survey of public health serotyping. *Epidemiol. Infect.* **129:**1–8.

Herwaldt, B. 2000. *Cyclospora cayetanensis*: a review, focusing on the outbreaks of cyclosporiasis in the 1990s. *Clin. Infect. Dis.* **31:**1040–1057.

Herwaldt, B., M. Ackers, and the Cyclospora Working Group. 1997. An outbreak in 1996 of cyclosporiasis associated with imported raspberries. *N. Engl. J. Med.* **336:**1548–1556.

Horby, P., S. O'Brien, G. Adak, C. Graham, J. Hawker, P. Hunter, C. Lane, A. Lawson, R. Mitchell, M. Reacher, E. Threlfall, L. Ward, and PHLS Outbreak Investigation Team. 2003. A national outbreak of multi-resistant *Salmonella enterica* serovar Typhimurium definitive phage type (DT) 104 associated with consumption of lettuce. *Epidemiol. Infect.* **130:**169–178.

Health Protection Agency (HPA). 2004a. *Salmonella* enteritidis non-phage type 4 infections in England and Wales: 2000 to 2004—report from a multi-agency national outbreak control team. *Communicable Dis. Rep. CDR Wkly.* **14**:news [Online.] http://www.hpa.org.uk/cdr/archives/2004/cdr4204.pdf. Accessed 24 July 2007.

Health Protection Agency (HPA). 2004b. Controlling the national outbreaks of *Salmonella* Enteritidis non-phage type 4 infection in England & Wales 2002–2004. Report from a multi-Agency National Outbreak Control Team. [Online.] http://www.hpa.org.uk/infections/topics_az/salmonella/OCT_position.pdf. Accessed 24 July 2007.

Isaacs, S., J. Aramini, B. Ciebin, J. Farrar, R. Ahmed, D. Middleton, A. Chandran, L. Harris, M. Howes, E. Chan, A. Pichette, K. Campbell, A. Gupta, L. Lior, M. Pearce, C. Clark, F. Rodgers, F. Jamieson, I. Brophy, A. Ellis, and *Salmonella* Enteritidis PT30 Outbreak Investigation Working Group. 2005. An international outbreak of salmonellosis associated with raw almonds contaminated with a rare phage type of *Salmonella* enteritidis. *J. Food Protect.* **68**:191–198.

Jerardo, A. 2003. Import share of U.S. food consumption stable at 11 percent. [Online.] http://www.ers.usda.gov/Publications/FAU/July02/FAU6601. Accessed 1 February 2007.

Kapperud, G., L. Rørvik, V. Hasseltvedt, E. Høiby, B. Iversen, K. Staveland, G. Johnsen, J. Leitao, H. Herikstad, Y. Andersson, G. Langeland, B. Gondrosen, and J. Lassen. 1995. Outbreak of *Shigella sonnei* infection traced to imported iceberg lettuce. *J. Clin. Microbiol.* **33**:609–614.

Killalea, D., L. Ward, D. Roberts, J. de Louvois, F. Sufi, J. Stuart, P. Wall, M. Susman, M. Schwieger, P. Sanderson, I. Fisher, P. Mead, O. Gill, C. Bartlett, and B. Rowe. 1996. International epidemiological and microbiological study of outbreak of *Salmonella agona* infection from a ready to eat savoury snack. I. England and Wales and the United States. *Br. Med. J. (Clin. Res. Ed.)* **313**:1105–1107.

Kiple, K. 2007. *A Moveable Feast: Ten Millenia of Food Globalization.* Cambridge University Press, New York, N.Y.

Kirk, M. 2001. *Salmonella* Stanley, peanuts—Australia: recall. ProMed Mail, 11 September 2001. Archive no. 20010911.2189 [Online.] http://www.promedmail.org. Accessed 4 January 2007.

Kirk, M., C. Little, M. Lem, M. Fyfe, D. Genobile, A. Tan, J. Threlfall, A. Paccagnella, D. Lightfoot, H. Lyi, L. McIntyre, L. Ward, D. Brown, S. Surnam, and I. Fisher. 2004. An outbreak due to peanuts in their shell caused by *Salmonella enterica* serotypes Stanley and Newport—sharing molecular information to solve international outbreaks. *Epidemiol. Infect.* **132**:571–577.

Levine, W., R. Bennett, Y. Choi, K. Henning, J. Rager, K. Hendricks, D. Hopkins, R. Gunn, and P. Griffin. 1996. Staphylococcal food poisoning caused by imported canned mushrooms. *J. Infect. Dis.* **173**:1263–1267.

Little, C., S. Surman-Lee, M. Greenwood, F. Bolton, R. Elson, R. Mitchell, G. Nichols, and S. Sagoo. 2007. Public health investigations of *Salmonella* Enteritidis in catering raw shell eggs, 2002–2004. *Lett. Appl. Microbiol.* **44**:595–601.

Lynch, M., J. Painter, R. Woodruff, and C. Braden. 2006. Surveillance for foodborne disease outbreaks—United States, 1998–2002. *Morb. Mortal. Wkly. Rep. Surveill. Summ.* **55**:1–42.

Lyytikainen, O., J. Koort, L. Ward, R. Schildt, P. Ruutu, E. Japisson, M. Timonen, and A. Siitonen. 2000. Molecular epidemiology of an outbreak caused by *Salmonella enterica* serovar Newport in Finland and the United Kingdom. *Epidemiol. Infect.* **124**:185–192.

MacDonald, K., M. Eidson, C. Strohmeyer, M. Levy, J. Wells, N. Puhr, K. Wachsmuth, N. Hargrett, and M. Cohen. 1985. A multistate outbreak of gastrointestinal illness caused by enterotoxigenic *Escherichia coli* in imported semisoft cheese. *J. Infect. Dis.* 151:716–720.

MacDonald, P., R. Whitwam, J. Boggs, J. MacCormack, K. Anderson, J. Reardon, J. Saah, L. Graves, S. Hunter, and J. Sobel. 2005. Outbreak of listeriosis among Mexican immigrants as a result of consumption of illicitly produced Mexican-style cheese. *Clin. Infect. Dis.* 40:677–682.

Mahon, B., A. Ponka, W. Hall, K. Komatsu, S. Dietrich, A. Siitone, G. Cage, P. Hayes, M. Lambert-Fair, N. Bean, P. Griffin, and L. Slutsker. 1997. An international outbreak of *Salmonella* infections caused by alfalfa sprouts grown from contaminated seeds. *J. Infect. Dis.* 175:876–882.

McKee, M., and E. Steyger. 1997. When can the European Union restrict trade on grounds of public health? *J. Public Health Med.* 19:85–86.

Mead, P., and E. Mintz. 1996. Ethnic eating: foodborne disease in the global village. *Infect. Dis. Clin. Pract.* 5:319–323.

Mead, P., L. Slutsker, V. Dietz, L. McCaig, J. Bresee, C. Shapiro, P. Griffin, and R. Tauxe. 1999. Food-related illness and death in the United States. *Emerg. Infect. Dis.* 5:607–625.

Mintz, E., J. Bartram, P. Lochery, and M. Wegelin. 2001. Not just a drop in the bucket: Expanding access to point-of-use water treatment systems. *Am. J. Public Health* 91:1565–1570.

Moe, C., and R. Rheingans. 2006. Global challenges in water, sanitation and health. *J. Water Health* 4:41–57.

Moren, A., M. Rowland, F. Van Loock, and J. Giesecke. 1996. The European Programme for Intervention Epidemiology Training. *Eurosurveillance* 1:30–31. [Online.] www.epiet.org (see for current membership). Accessed 24 October 2007.

Naimi, T., J. Wicklund, S. Olsen, G. Krause, J. Wells, J. Bartkus, D. Boxrus, M. Sullivan, H. Kassenborg, J. Besser, E. Mintz, M. Osterholm, and C. Hedberg. 2003. Concurrent outbreaks of *Shigella sonnei* and enterotoxigenic *Escherichia coli* associated with parsley: implications for surveillance and control of foodborne illness. *J. Food Prot.* 66:535–541.

O'Brien, S., I. Gillespie, M. Sivanesan, R. Elson, C. Hughes, and G. Adak. 2006. Publication bias in foodborne outbreaks of infectious intestinal disease and its implications for evidence-based food policy. England and Wales 1992–2003. *Epidemiol. Infect.* 134:667–74.

O'Grady, K., J. Powling, A. Tan, M. Vulcanis, D. Lightfoot, and J. Gregory. 2001. *Salmonella* Typhimurium Dt104—Australia, Europe. ProMed Mail, 22 August 2001, Archive no. 20010822.1980 [Online.] http://www.promedmail.org. Accessed 4 January 2007.

OzFoodNet Working Group. 2005. Reported foodborne illness and gastroenteritis in Australia: annual report of the OzfoodNet network, 2004. *Commun. Dis. Intel.* 29:165–192.

OzFoodNet Working Group. 2006. Burden and causes of foodborne disease in Australia: Annual report of the OzFoodNet network, 2005. *Commun. Dis. Intel.* 30:278–300.

Penteado, A., B. Eblen, and A. Miller. 2004. Evidence of *Salmonella* internalization into fresh mangos during simulated postharvest insect disinfestation procedures. *J. Food Prot.* 67:181–184.

Rushing, J. W., F. J. Angulo, and L. R. Beuchat. 1996. Implementation of a HACCP program in a commercial fresh-market tomato packinghouse: A model for the industry. *Dairy Food Environ. Sanit.* 16:549–553.

Sepulveda, J., F. Bustreao, R. Tapia, J. Rivera, R. Lozano, G. Olaiz, V. Partida, L. Garcia-Garcia, and J. Valdespino. 2006a. Improvement of child survival in Mexico: the diagonal approach. *Lancet* **368**:2017–2027.

Sepulveda, J., J. Valdespino, and L. Garcia-Garcia. 2006b. Cholera in Mexico: the paradoxical benefits of the last pandemic. *Int. J. Infect. Dis.* **10**:4–13.

Shohat, T., M. Green, D. Merom, O. Gill, A. Reisfeld, A. Matas, D. Blau, N. Gal, and P. Slater. 1996. International epidemiological and microbiological study of outbreak of *Salmonella agona* infection from a ready to eat savoury snack. II. Israel. *Br. Med. J. (Clini. Res. Ed.)* **313**:1107–1109.

Sivapalasingam, S., E. Barrett, A. Kimura, M. Van Duyne, W. De Witt, M. Ying, A. Frisch, Q. Phan, E. Gould, P. Shillam, V. Reddy, T. Cooper, M. Hoekstra, C. Higgins, J. Sanders, R. Tauxe, and L. Slutsker. 2003. A multistate outbreak of *Salmonella enterica* serotype Newport infections linked to mango consumption: Impact of a water-dip disinfestation technology. *Clin. Infect. Dis.* **37**:1585–1590.

Swaminathan, B., P. Gerner-Smidt, L.-K. Ng, S. Lukinmaa, K.-M. Kam, S. Rolando, E. Perez Gutierrez, and N. Binsztein. 2006. Building PulseNet International: An interconnected system of laboratory networks to facilitate timely public health recognition and response to foodborne disease outbreaks and emerging foodborne diseases. *Foodborne Pathog. Dis.* **3**:36–50.

Tanner, C. 1997. Principles of Australian quarantine. *Aust. J. Agric. Res. Econ.* **41**:541–558.

Taormina, P., L. Beuchat, and L. Slutsker. 1999. Infections associated with eating seed sprouts: an international concern. *Emerg. Infect. Dis.* **5**:626–634.

Tauxe, R. V. 2006. Molecular subtyping and the transformation of public health. *Foodborne Pathog. Dis.* **3**:4–8.

Tauxe, R., and J. Hughes. 1996. International investigation of outbreaks of foodborne disease. *Br. Med. J. (Clin. Res. Ed.)* **313**:1093–1094.

Taylor, D., and S. Woodgate. 1997. Bovine spongiform encephalopathy: the causal role of ruminant-derived protein in cattle diets. *Rev. Sci. Technol.* **16**:187–198.

Taylor, J. L., J. Tuttle, T. Pramukul, K. O'Brien, T. J. Barrett, B. Jolbitado, Y. L. Lim, D. Vugia, J. G. J. Morris, R. V. Tauxe, and D. M. Dwyer. 1993. An outbreak of cholera in Maryland associated with imported commercial frozen fresh coconut milk. *J. Infect. Dis.* **167**:1330–1335.

Ueseqi, A., and L. Harris. 2006. Growth of *Salmonella* Enteritidis phage type 30 in almond hull and shell slurries and survival in drying almond hulls. *J. Food Prot.* **69**:712–718.

Unicomb, L., M. Kirk, G. Hogg, P. Jelfs, G. Simmons, J. Gregory, C. Nicol, and Eurosurveillance Editorial Team. 2003. *Salmonella* Montevideo in sesame seed-based products imported into Australia and New Zealand may have implications for Europe and elsewhere. *Eurosurveillance* **7**:030918.

Unicomb, L., G. Simmons, T. Merritt, J. Gregory, C. Nicol, P. Jelfs, M. Kirk, A. Tan, R. Thomson, J. Adamopoulos, C. Little, A. Currie, and C. Dalton. 2005. Sesame seed products contaminated with *Salmonella*: three outbreaks associated with tahini. *Epidemiol. Infect.* **133**:1065–1072.

Vaillant, V., S. Haeghebaert, J.-C. Desenclos, P. Bouvet, F. Grimont, P. Grimont, and A. Burnens. 1996. Outbreak of *Salmonella dublin* infection in France, November-December 1995. *Eurosurveillance* **1**:9–10.

Villar, R., M. Macek, S. Simons, P. Hayes, M. Goldoft, J. Lewis, L. Rowan, D. Hursh, M. Patnode, and P. Mead. 1999. Investigation of multidrug-resistant *Salmonella* serotype Typhimurium DT104 infections linked to raw-milk cheese in Washington State. *J. Am. Med. Assoc.* 281:1811–1816.

Voetsch, A., T. Van Gilder, F. Angulo, M. Farley, S. Shallow, R. Marcus, R. Cieslak, V. C. Deneen, and R. Tauxe. 2004. FoodNet estimate of the burden of illness caused by nonty-phoidal *Salmonella* infections in the United States. *Clin. Infect. Dis.* 38:S127–S134.

Webby, R., K. Carville, M. Kirk, G. Greening, R. Ratcliff, S. Crerar, K. Dempsey, M. Sarna, R. Stafford, M. Patel, and G. Hall. 2007. Internationally distributed frozen oyster meat caus-ing multiple outbreaks of norovirus infection in Australia. *Clin. Infect. Dis.* 44:1026–1031.

Weber, J. T., E. D. Mintz, R. Canizares, A. Semiglia, I. Gomez, R. Sempertegui, A. Davila, K. D. Greene, N. D. Puhr, D. N. Cameron, F. C. Tenover, T. J. Barrett, N. H. Bean, C. Ivey, R. V. Tauxe, and P. A. Blake. 1994. Epidemic cholera in Ecuador: Multidrug-resistance and transmission by water and seafood. *Epidemiol. Infect.* 112:1–11.

Wheeler, C., T. Vogt, G. Armstrong, G. Vaughan, A. Weltman, O. Nainan, V. Dato, G. Xia, K. Waller, J. Amon, T. Lee, A. Highbaugh-Battle, C. Hembree, S. Evenson, M. Ruta, I. Williams, A. Fiore, and B. Bell. 2005. An outbreak of Hepatitis A associated with green onions. *N. Engl. J. Med.* 353:890–897.

World Health Organization (WHO). 2005. WHA 58.3 Revision of the International Health Regulations. [Online.] http://www.who.int/csr/ihr/en/. Accessed 21 June 2007.

Wilesmith, J., J. Ryan, W. Hueston, and L. Hoinville. 1992. Bovine spongiform en-cephalopathy: epidemiological features 1985 to 1990. *Vet. Rec.* 130:90–94.

Winthrop, K., M. Palumbo, J. Farrar, J. Mohle-Boetani, S. Abbott, M. Beatty, G. Inami, and, S. Werner. 2003. Alfalfa sprouts and *Salmonella* Kottbus infection: a multistate outbreak following inadequate seed disinfection with heat and chlorine. *J. Food Prot.* 66:13–17.

Imported Foods: Microbiological Issues and Challenges
Edited by Michael P. Doyle and Marilyn C. Erickson
© 2008 ASM Press, Washington, DC

Animal and Human Waste as Vehicles for Cross-Contamination of Imported Foods

4

Charles P. Gerba and Christopher A. Scott

INTRODUCTION

This chapter focuses on animal and human wastes as potential sources of contamination of imported foods. Global trade of food, in particular, fresh fruit, vegetables, and seafood, has risen dramatically since the 1980s (Huang, 2004); however, not without concern for food safety. Most food exports to the United States (≈80%) are regulated by the U.S. Food and Drug Administration. Produce has been a vehicle for the importation of infectious disease into the United States, and the number of outbreaks associated with produce in the United States is on the increase (Sivapalasingam et al., 2004). Public awareness of food safety and potential health risks has increased dramatically, largely based on recent reports of contaminated seafood imported to the United States and pathogen outbreaks associated with domestically produced fresh-cut leafy greens; these public health issues have led to import bans or lawsuits against producers.

The amount of animal and human feces generated on a worldwide basis is enormous. Humans generate 900 billion pounds a year and livestock more than 24 trillion pounds (World Bank, 2005). Exposure to feces may occur by the use of wastes for fertilization of a crop, the use of contaminated water in the production of a product (e.g., irrigation water or water used to produce shellfish), or human handling during harvesting or in the processing of a product. The use of human wastes or wastewater for produce production in developed countries is not allowed or is strictly regulated because of the potential for transmission of enteric pathogens. Although use of highly treated

CHARLES P. GERBA, Department of Soil, Water and Environmental Science, University of Arizona, Tucson, AZ 85721. CHRISTOPHER A. SCOTT, Department of Geography and Regional Development, and, Udall Center for Studies in Public Policy, University of Arizona, Tucson, AZ 85721.

domestic wastewater for food crops to be eaten raw is allowed in some states in the United States, it is seldom practiced because of potential health hazards. The same is true for sewage sludge (biosolids). In the United States, more than half of all biosolids is recycled by application to agricultural land. However, almost the entire amount goes to land where food crops are not grown. Animal wastes have been used for a long time in food crop production; however, such waste is often composted before utilization. In contrast, irrigation with sewage in the developing world is commonly practiced, especially in water-scarce areas (Scott et al., 2004). Direct use of human wastes as a fertilizer is a centuries-old practice in China. Also, much of the world still discharges raw or incompletely treated sewage into rivers and other bodies of water that could be used for irrigation, washing of produce, or growing shellfish.

Contamination of food crops may also occur during human handling of crops before harvest (e.g., turning of cantaloupe to prevent damage to the rind), during harvesting, and during processing. Use of contaminated water during washing or cooling may be an additional source of exposure.

PATHOGENS TRANSMITTED

Domestic animals and humans share many common enteric bacteria, protozoan and worms, which can be easily transmitted between species. In contrast, enteric viruses are usually host specific and are not transmitted between species. The only enteric virus known to be transmitted between nonprimates is hepatitis E virus. The virus infects both man and swine and has been shown to be food-borne (Li et al., 2005). Some respiratory viruses have the potential to be transmitted by water via human or animal waste. For example, influenza virus is present in animal wastes (e.g., it is excreted in the feces of birds) and has the potential to be transmitted by foods but has not yet received serious investigation. Table 1 lists enteric zoonotic pathogens that potentially can be transmitted through the environment. Emerging pathogens continue to be recognized, and new ones are added to the list of food-borne risk factors almost annually.

Only some enteric bacteria, not viruses or protozoa, have the potential to grow in the environment or in/on foods. However, even the enteric bacteria have a limited existence in the environment outside of an animal. Of the enteric bacteria, *Salmonella*, *Campylobacter*, and *Escherichia coli* O157:H7 present the greatest problem because they infect a wide variety of animals besides humans. *Shigella* spp. are only known to infect humans and some primates; thus, they are usually only present in human wastes.

Cryptosporidium and *Giardia* present the greatest problem among the protozoan parasites and are a common cause of gastroenteritis worldwide. Cattle are

Table 1 Enteric bacteria, protozoa, microsporidia, and helminthes that infect animals and humans

Group	Pathogen	Disease or condition
Bacteria	*Salmonella*	Gastroenteritis, typhoid fever
	Campylobacter	Gastroenteritis
	Vibrio cholerae	Gastroenteritis
	Yersinia enterocolitica	Gastroenteritis
	Escherichia coli (certain strains)	Gastroenteritis
Protozoa	*Giardia intestinalis*	Gastroenteritis
	Cryptosporidium	Gastroenteritis
Microsporidia	*Enterocytozoon bieneusi*	Gastroenteritis in immunocompromised person
Helminthes	*Ascaris lumbricoides*	Ascariasis
	Necater americanus	Hookworm
	Tania saginata	Tapeworm (from beef)

a major source of *Cryptosporidium* worldwide, while humans appear to be the major source of *Giardia intestinalis*. Foods imported into the United States and other developed countries have been a source of the protozoan parasite *Cyclospora* (Ho et al., 2002; Hoang et al., 2005). At present, it is not known whether this parasite is limited to humans or has an animal reservoir. *Entamoeba histolytica* is a protozoan parasite that only infects humans (Table 2), while microsporidia only appear to cause serious infections in the immuno-compromised; their role in infection of healthy persons is currently unknown.

Human enteric noroviruses are believed to be the major cause of food-borne illnesses in developed countries (Gerba and Kayed, 2003). Noroviruses cause an estimated 67% of all food-borne illnesses each year in the United States (Mead et al., 1999). Hepatitis A virus has been the cause of shellfish- and

Table 2 Enteric viruses, bacteria, and protozoa that infect only humans

Group	Pathogen(s)	Disease or condition
Viruses	Enteroviruses	Meningitis, paralysis, rash, fever, heart disease, respiratory disease
	Hepatitis A	Hepatitis
	Rotaviruses	Gastroenteritis
	Noroviruses	Gastroenteritis
	Adenoviruses	Gastroenteritis, respiratory illness, eye infections
	Astroviruses	Gastroenteritis
Bacteria	*Shigella* spp.	Gastroenteritis
Parasites	*Entamoeba histolytica*	Gastroenteritis

Table 3 Occurrence of enteric pathogens in human feces in the United States[a]

Pathogen	Incidence (%)	Concn in stool (per gram)
Enteroviruses	10	10^3-10^8
Hepatitis A	0.1	10^8
Rotavirus	10–29	$10^{10}-10^{12}$
Giardia	3.8	10^6
Cryptosporidium	0.6–20	10^6-10^7

[a] Source: Gerba, 2000.

imported produce-associated outbreaks in the United States (Sanchez et al., 2007). Rotavirus is the most common cause of childhood gastroenteritis worldwide, but large water-borne (Gerba et al., 1996) and food-borne outbreaks (Centers for Disease Control and Prevention [CDC], 2000) also have been documented in adults. Enteroviruses have been less commonly associated with food- and water-borne outbreaks, probably because the wide variety of illnesses they are capable of causing (Table 2) makes it difficult to identify them as the causative agents.

OCCURRENCE OF PATHOGENS

The occurrence of enteric pathogens in the environment is a reflection of the incidence of infection by the pathogen in a population. In general, the greater the incidence of infection, the greater the occurrence of a particular pathogen in the environment. Enteric pathogens can be excreted in large numbers in the feces of infected humans and animals (Table 3), whether or not the infected individual exhibits the symptoms of clinical illness. Infected individuals may continue to excrete pathogens for many weeks to months after infection—long after clinical signs of illness have passed. The number of asymptomatic infected individuals will vary depending on the virulence of the organism, immune state of the individual, species of host, and age. For example, in the case of hepatitis A virus, only 5% of infected children develop clinical disease, whereas 95% of the nonimmune adults will develop clinical illness (Cuthbert, 2001).

Pathogen concentrations may also vary with the type of wastewater. For example, various studies have indicated that *E. coli* O157:H7 is commonly present in animal and human wastewaters at levels of 10 to 100/100 ml of municipal sewage and 100 to 1,000/100 ml of animal wastewaters and wastewaters from slaughterhouses (Muniesa et al., 2006).

Human Wastes

Feces

Humans excrete between 150 to 400 g of feces per day, depending on diet (Feachem et al., 1983). Infected persons may excrete large numbers of pathogens in their feces (Table 3). Viruses are generally excreted in greater concentrations than bacteria and parasites. Rotaviruses are excreted in concentrations as great as 10^{12}/g or more than 10^{14}/day. The incidence of persons excreting viruses in the United States varies greatly with age group and season, but 10 to 40% of children may be excreting viruses at any point in time (Gerba, 2000). The prevalence of enteric pathogen excretion is much greater in developing countries because of much greater infection rates. Lack of adequate sanitation facilities and poor hygienic practices, in general, in developing countries compound the risk of exposure and reinfection, which are already high because of elevated pathogen concentrations in excreta.

Sewage

Because some persons in a population are always infected, enteric pathogens are always present in raw domestic sewage. The concentration will vary depending on several important factors:

- Incidence of infection within the community
- Public health status within the community
- Season
- Social-economic conditions (in developing countries this may reflect access to pathogen-contaminated water and sanitation, whereas in developed countries this may be a reflection of access to health care)
- Per capita water use

Because of these factors, the concentration of pathogens in raw sewage is much greater in developing countries, a factor that is often overlooked when assessing the risks from the use of wastewater in irrigation of food crops (Blumenthal et al., 2000, 2001; Ensink et al., 2004). The concentrations of pathogens detected in raw sewage are shown in Table 4. In developing countries, the concentration of enteric pathogens is 100 times that reported in developed countries, e.g., 1,000 versus 100,000 per liter (Buras, 1976). In Brazil, the average concentration of rotavirus is 11,300 per liter and as great as 26,600 per liter (Oragui, 2003). In contrast, the concentration of rotavirus in domestic sewage in the United States ranges from 1 to 321 per liter (Hejkal et al., 1984). This is an important consideration because the ratio of fecal indicator bacteria (e.g., E. coli) may be much different in a developing country versus a developed country. Thus, the risk of exposure to enteric pathogens with a given concentration of fecal indicator bacteria is greater in a developing country than in

Table 4 Concentration of enteric pathogens in domestic sewage in developed versus developing countries[a]

Microbe	Concn per liter	
	Developed countries	Developing countries
Enteric viruses	10^3	10^4–10^5
Giardia	10^1–10^3	10^2–10^5
Cryptosporidium	10–10^2	10–10^3

[a]Source: Buras, 1976; Feachem et al., 1983; Jimenez et al., 2002; Smith and Grimason, 2003.

developed countries. This is true of environmental health conditions, in general, in developing countries, i.e., elevated pathogen concentrations not just in sewage, but in water supplies, on fruits and vegetables, and in soils (Carr et al., 2004)—a factor recognized in the *Stockholm Framework* that informs the 2006 World Health Organization wastewater guidelines (WHO, 2006). The *Stockholm Framework* seeks to manage health risks from all water-related microbial exposures and encourages flexibility to adapt and apply guidelines based on local conditions (social, cultural, economic, and environmental).

While treatment of sewage is primarily designed to reduce the amount of biodegradable organic matter, major reductions of pathogens also occur. Because of their large size, helminthes are reduced in large numbers just by settling. Activated sludge is the most common method used for the treatment of domestic sewage in developed countries and is increasing in use in developing countries. In this process, bacteria are most effectively removed, followed by enteric viruses and protozoan parasites (Table 5). Chlorination as commonly practiced can result in significant reduction of bacterial

Table 5 Removal and estimated concentration of enteric pathogens after treatment of domestic sewage[a]

Parameter	Enteric viruses	*Salmonella*	*Giardia*
Concn in raw sewage per liter	1,000–100,000	5,000–200,000	9,000–200,000
Primary treatment			
% removal[a]	50–98	96–99.8	27–64
No. remaining/ liter	170–50,000	160–3,360	72,000–146,000
Activated sludge			
% removal	53–99.9	98–99.996	45–96.7
No. remaining/ liter	8–47,000	3–1075	6,480–109,500
After chlorination			
% removal	90–96	99.9–99.99	0–20
No. remaining/ liter	1–4,700	$3 \times 10^{-4} - 1$	5,184–87,600

[a] Source: Maier et al., 2000.

pathogens but has only a nominal effect on the more resistant viruses and protozoa. Ultraviolet light disinfection is very effective against enteric bacteria and protozoa, but some viruses (adenoviruses) are extremely resistant (Meng and Gerba, 1996).

Waste stabilization ponds (oxidation ditches, stabilization lagoons) are often used to treat wastes in the developing world and are still used for small systems in the United States. They are commonly used in tropical and subtropical regions, where the warm climate results in increased performance and lower land area requirements. They have been promoted as a low-cost method of sewage treatment and means to reduce the concentration of pathogens. However, performance is based on residence time, climatic conditions (amount of sunlight), and minimization of short circuiting. Ponds are especially good at reducing the concentrations of helminthes and protozoan pathogens because they are large enough to settle in large numbers (Mara, 2002). In properly operated ponds, reduction of 99.9% or more of the bacteria and viral pathogens can be achieved (Mara, 2002; Oragui, 2003). Even well-operated ponds do not produce pathogen-free wastewater, and the wastewater is of lower quality than that produced by activated sludge sewage treatment plants.

Activated sludge treatment generates large amounts of sludge or biosolids, which must be either disposed of or recycled. Biosolids contain substantial levels of enteric pathogens. The larger helminthes ova settle out during activated sludge treatment and the viruses tend to adsorb to the biological floc that forms during this process. More than half of the biosolids generated in the United States is recycled through application to agricultural land. Sewage sludge is classified into two types, depending on the concentration of pathogens. Class B biosolids contain detectable levels of pathogens; food crops that may be eaten raw cannot be grown on land to which Class B solids have been applied for two years to allow for the die-off of ascaris ova. Class A biosolids are those that receive additional treatment to eliminate pathogens and can be used on food crops. Such processes include thermophilic anaerobic digestion, heating to high temperatures, composting, prolonged storage, and treatment at high pH (pH 12). These processes must reduce the fecal coliform levels to less than 1,000/g of dry solids, and no detectable enteric viruses, *Salmonella,* or viable *Ascaris* ova in 4 g of dry solids (Environmental Protection Agency [USEPA], 1993). Most developed countries have similar restrictions for the use of sewage sludge. Activated sludge treatment is becoming more common in some developing countries, and it has many benefits to agricultural production. However, the concentration of pathogens in sewage sludge generated in developing countries is much greater than in sewage sludge from developed countries (Table 6; Jimenez et al, 2002). Hence, sewage

Table 6 Concentration of enteric pathogens in untreated (raw) sewage sludge (biosolids) in the United States and Mexico[a,b]

Type	Microbe	United States	Mexico
Virus	Enteric viruses	300	No data
Bacteria	*Salmonella*	2,800	$10^6–10^8$
Protozoa	*Giardia*	10–1,000	$10^2–10^4$
Helminthes	*Ascaris*	<1–9.7	66–136
	Toxocara	1–5	0.3–1.2

[a]Concentration per gram of dried solids.
[b]Source: Straub et al., 1993; Jimenez et al., 2002.

sludge from developing countries will require additional treatment before it can be safely reused.

Animal Wastes

Grazing systems are estimated to cover 50% of all the agricultural land in the world (Gannon et al., 2004b). Hence, it is not surprising that zoonotic pathogens occur in most areas of the world where there is agricultural production. Animal wastes and effluents from farming operations, including manures and slurries, are frequently applied to increase the soil fertility of cropland. Animal manures have been used for centuries as a fertilizer and soil enhancer. Enteric bacterial pathogens have been the microbes most commonly associated with produce outbreaks, although protozoan parasites have also been responsible for produce-associated outbreaks. While hepatitis E virus is shed in feces by pigs (Kasorndorkbua et al., 2005) and can be transmitted through raw or uncooked meats (Li et al., 2005), other human enteric viruses are not currently considered to be zoonotic. The recent finding of human norovirus genotypes in swine and cattle feces, on the other hand, suggests that animal feces could be involved in their transmission (Mattison et al., 2007). Like human feces, animal wastes can contain high concentrations of human enteric pathogens ($>10^6$/g, Tables 7 and 8) (Gannon et al., 2004b). Young animals are often more likely than adult animals to shed pathogens, and at higher levels. For example, *E. coli* O157:H7 is shed in calves at concentrations from 10^2 to 10^6 per gram of feces, but less commonly isolated from adults (Gannon et al., 2004b). However, recent research suggests that a small number (2 to 7%) of the adult cattle population may be supershedders, excreting large numbers of *E. coli* O157:H7 (Omisakin et al., 2003; Matthews et al., 2006). *Cryptosporidium parvum*, which also infects humans, in England was found to infect 50% of farmed calves and was excreted at 10^5/g (Sturdee et al., 2003). The relative occurrence of human enteric pathogens in the feces of animals in the developing world appears to have received little attention.

Table 7 Percentage of Great Britain livestock manures contaminated with zoonotic microbes[a]

| | % Contamination | | | | | | | |
| | Cattle | | Swine | | Poultry | | Sheep | |
Pathogen	Fresh[b]	Stored[c]	Fresh	Stored	Fresh	Stored	Fresh	Stored
E. coli O157:H7	13.2	9.1	11.9	15.5	ND[d]	ND	20.8	22.2
Salmonella	7.7	10.0	7.9	5.2	17.9	11.5	8.3	11.1
Listeria	29.8	31.0	19.8	19.0	19.4	15.4	29.2	44.4
Campylobacter	12.8	9.8	13.5	10.3	19.4	7.7	20.8	11.1

[a]Source: Hutchison et al., 2004.
[b]Fresh, collected from location in which deposited
[c]Stored, collected from lagoon or farm yard manure heap
[d]ND, not determined

The closer association of humans with animals in the developing world generates more opportunities for transfer of viruses between livestock and humans with closely related pathogens (Kang et al., 2005).

Properly treated manure is generally a safe fertilizer, but improperly treated manure may contain pathogens that can contaminate fresh produce. The high temperature and production of ammonia play a critical role in the

Table 8 Cell numbers of zoonotic pathogens in British livestock manures[a]

| | No. of cells (CFU/g) of pathogen | | | | | | | |
| | Cattle | | Swine | | Poultry | | Sheep | |
Pathogen	Fresh[b]	Stored[c]	Fresh	Stored	Fresh	Stored	Fresh	Stored
E. coli O157								
Geo mean	1×10^3	3×10^2	4×10^3	1×10^3	ND[d]	ND	8×10^2	3×10^2
Max	3×10^8	8×10^4	8×10^5	2×10^4	ND	ND	5×10^4	5×10^3
Salmonella								
Geo mean	2×10^3	3×10^3	6×10^2	6×10^2	2×10^2	4×10^3	7×10^2	6×10^3
Max	6×10^5	7×10^6	8×10^4	2×10^3	2×10^4	8×10^3	2×10^3	6×10^3
Listeria								
Geo mean	1×10^3	1×10^3	3×10^3	6×10^2	8×10^2	3×10^2	2×10^2	3×10^2
Max	6×10^5	7×10^6	8×10^4	2×10^3	2×10^4	8×10^3	2×10^3	6×10^3
Campylobacter								
Geo mean	3×10^2	5×10^2	3×10^2	2×10^3	3×10^2	6×10^2	4×10^2	1×10^2
Max	2×10^5	2×10^5	2×10^4	1×10^5	3×10^4	9×10^2	2×10^3	1×10^2

[a]Source: Hutchison et al., 2000.
[b]Fresh, collected from location in which deposited.
[c]Stored, collected from lagoon or farm yard manure heap.
[d]ND, not determined.

reduction of enteric pathogens during composting (Haug, 1993). However, proper composting and subsequent storage conditions are critical to the destruction of pathogens and control of regrowth of enteric pathogens (Zaleski et al., 2005).

Composting systems are usually divided into three categories: windrow, static pile, and in-vessel. Usually a bulking agent such as wood chips is added to create air spaces during composting processes (Haug, 1993). In the windrow system, the manure is composted in long rows aerated by convective air movement. The rows are turned periodically by mechanical means (specialized machinery or dump loaders, which may lead to recontamination). Turning exposes the organic matter to oxygen and ensures that all of the material is treated to a high enough temperature to kill pathogens. In a static pile (or forced aeration), piles of manure may be aerated by using a forced aeration system, which is installed under the piles to maintain a minimum oxygen level throughout the compost pile. In-vessel composting (also known as enclosed-reactor composting) takes place in partially or completely enclosed containers under controlled environmental conditions. In windrow composting, it is necessary for temperatures to be at least 131°F (55°C) for 15 days and for the piles to be turned over at least five times to ensure pathogen destruction. In static aerated piles or in-vessel composting, this temperature must be maintained for at least 3 days (USEPA, 1993).

Once composting is completed, there is the potential for regrowth of surviving *Salmonella* and *E. coli* if the moisture increases or contamination occurs from animals (e.g., birds) (Zaleski et al., 2005). Temperatures above 28°C have been favorable to the growth of *Salmonella* in compost. However, once compost was added to soil at typical agronomic rates (3 tons per acre), growth did not occur even when the *Salmonella* was added to the soil/compost in large numbers, probably because of competition or predation from native microflora (Campo et al., 2007).

Surface Waters

Any freshwater surface source is likely to contain enteric pathogens periodically. The risk of pathogen exposure depends on the use of water and the pathogen concentration in the water at the time of utilization. The greatest levels of pathogens occur in water sources receiving discharges of untreated sewage and in watersheds with intensive levels of animal production. In the United States, sewage discharges are usually disinfected. Storm water runoff from agricultural land and septic tanks (referred to as nonpoint sources) is the largest contributor of pathogens to surface water. In Europe and much of the rest of the world, treated sewage discharges are not usually disinfected. Hence, human enteric viruses can be isolated from almost every major river in Europe

(Bosch et al., 2006). Untreated or partially treated sewage is still discharged into surface waters in much of the world, and isolation of enteric pathogens is common in surface waters in developing countries.

The greatest loading of pathogens occurs in surface waters after rainfall events, although concentrations may be elevated during low flow when the proportion of wastewater to natural runoff is highest. During runoff events, accumulated feces are washed into near-by streams or collection systems forcing release of partially or untreated sewage from wastewater treatment plants. Studies of water-borne disease outbreaks in the United States and Canada have correlated above-average rainfall events with drinking water-associated disease outbreaks (Curriero et al., 2001; Thomas et al., 2006). For these reasons, gastroenteritis is usually more common in the rainy season in developing countries.

Use of contaminated water for pesticides may also serve as another mechanism of produce contamination. Guan et al. (2005) found that *Salmonella* and *E. coli* O157:H7 could grow in various pesticide solutions and contaminate tomato plants when they are applied as a spray.

Ground Waters

In general, enteric pathogens are far less common in groundwater because of the natural filtering mechanism of soil. However, every groundwater source is potentially susceptible to contamination. The construction of a well (or location of a spring), nature of the substrata, depth to groundwater, and rainfall can affect the microbiologic quality of the well water. Even in the United States, half of all drinking water-associated disease outbreaks reported annually are caused by contaminated groundwater (Reynolds et al., 2007). Pathogens enter groundwater from latrines, septic tank leach fields, land application of wastewater for irrigation, oxidation ponds, leaking sewer lines, and unlined landfills. Unprotected wells may allow for surface water to run into the well during storm events. In well-structured soils, helminthes ova, protozoa, and bacteria are easily filtered out during transport through the soil. However, in fractured limestone and clay soils, long-distance transport is possible. Viruses are more likely to contaminate groundwater and travel long distances (hundreds of meters) because of their small size. Microorganisms are biocolloids that behave as particulates in their transport in the subsurface; hence, their movement may be much different than solutes, depending on the nature of the subsurface matrix. Their transport in fractured substrata may be 100 times faster than solutes such as nitrates (McKay et al., 1993). Hence, lack of chemical contamination does not ensure the absence of enteric pathogens. The potential for groundwater contamination by enteric viruses should not be underestimated. In a survey of more than 400 utility drinking water wells across the United

States, enteric viruses were detected in 32%. Moreover, indicator bacteria were not found to be useful indicators of virus contamination (Abbaszadegan et al., 2003).

Irrigation Waters

Irrigation using surface and ground waters mirrors similar water quality and pathogen concentrations, as discussed in the preceding two sections. With growing urban populations that have increased sewerage collection, although often without sewage treatment, irrigation water used downstream of large urban centers has high pathogen loads (Scott et al., 2004). Efforts to undertake a global assessment of irrigation with wastewater (van der Hoek, 2004) have been difficult, given the lack of national-level data and the need to examine a very large number of localities.

Studies are limited on the occurrence of pathogens in irrigation water impacted by nondirect or purposeful reuse of sewage in both developed and developing countries. In a study of irrigation waters in several Central American countries and the United States, the protozoan parasites *Giardia*, *Cryptosporidium*, and microsporidia were detected (Thurston-Enriquez et al., 2002). *Giardia* concentrations were similar in almost all countries (60% of the samples), whereas *Cryptosporidium* concentrations were much greater in Central America. In a study in western Mexico, 48% of the surface irrigation waters were positive for *Cryptosporidium* oocysts and 50% for *Giardia* cysts (Chaidez et al., 2005). Both of these parasites have also been detected in irrigation water used for bean sprout irrigation in Norway (Robertson and Gjerde, 2001). *Salmonella* has been reported in 2 to 14% of the irrigation waters used for produce production in Nigeria (Okafo et al., 2003) and 23.5% of irrigation water used for cantaloupe production in Brazil (Espinoza-Medina et al., 2006). Both *E. coli* O157:H7 and *Salmonella* have been reported in irrigation waters in western Canada (Gannon et al., 2004a).

SURVIVAL OF PATHOGENS

Principal factors controlling the survival of enteric pathogens in the environment are temperature, moisture content, and exposure to sunlight. Knowledge of these factors enables the estimation of pathogen survival in the environment. Other factors influencing survival are type and strain of microbe, pH, organic matter, and antagonism by native microflora. Temperature is the major factor in all environments affecting survival. Moisture and desiccation are important factors in soils and on plant surfaces (Stine et al., 2005a). In surface waters, ultraviolet light in sunlight is a dominating factor. Regrowth of

Table 9 Die-off rates of enteric pathogens in the environment[a]

Type	Organism	Die-off ($\log_{10}$/day)
Viruses	Enteroviruses	0.01–0.2
	Hepatitis A	0.01–0.2
Bacteria	*Salmonella* spp.	1–7
	E. coli	0.23–0.46
Protozoa	*Giardia*	0.023–0.23
	Cryptosporidium	0.0057–0.046

[a]Source: Madema et al., 2003.

bacterial pathogens is possible in aquatic sediments, feces, manure, compost, and plant surfaces (Zaleski et al., 2005).

Most enteric pathogens can survive for at least a few days in the environment, but under the right conditions they may persist for years. Ascaris ova may survive for as long as two years in soil (Gerba and Smith, 2005). Enteric viruses and protozoan parasites are also capable of prolonged survival, especially at low temperatures (10°C or less), and may survive for many months (Table 9). Because of its resistance to inactivation at elevated temperatures, hepatitis A virus is likely among the longest surviving of the enteric viruses (Stine et al., 2005a). Enteric bacterial pathogens generally have the shortest survival time (days to weeks), although unlike the other enteric pathogens, they do have the ability to persist and increase in number under the right environmental conditions (largely temperature and nutrient dependent). As a general rule, survival of enteric pathogens is in the following order: soil > plants > water (Gerba and Smith, 2005; Stine et al., 2005b).

Soil

Enteric pathogen survival depends on both the type of microbe and the type of soil (Hurst et al., 1980). Die-off will be much greater in warm tropical conditions than in cool damp climates. Enteric viruses generally survive several weeks to months, depending on soil temperature. Helminthes, in particular, can survive up to years in the soil (Gerba and Smith, 2005).

Plants

Few studies exist on the survival of enteric pathogens on produce preharvest. (Fattal et al., 2004). *Salmonella* and *E. coli* may be capable of growth on produce surfaces under certain conditions (Stine et al., 2005a). Poliovirus was found to survive on lettuce after flood irrigation in outdoor plots (Tierney et al., 1977). The virus persisted on the lettuce for two months during the winter, but only 2 to 3 days in the summer. Stine et al. (2005a) found that hepatitis A virus and the feline calicivirus survived better than enteric bacteria on cantaloupe, lettuce, and bell peppers. Little decrease of hepatitis A virus occurred during the two-week study period. Survival of enteric pathogens on produce is influenced by a number of factors, including competition from native microflora, humidity, sunlight, and temperature (Nyeleti et al., 2004; Stine et al., 2005a; Cooley et al., 2006).

Little inactivation of enteric pathogens postharvest is expected if they are chilled or refrigerated. Studies on the survival of enteric viruses on produce postharvest indicate that little virus inactivation occurs because of the low temperatures of storage (Seymour and Appleton, 2001).

ROUTES OF CROSS-CONTAMINATION

Irrigation Water

The largest use of freshwater in the world is in agriculture, with more than 70% being used for irrigation. Approximately 240 million ha, 17% of the world's cropland, are irrigated, producing one-third of the world's food supply (Shannan, 1998). Nearly 70% of this area is in developing countries. Irrigation of food crops with untreated domestic sewage has long been associated with the transmission of infectious diseases. As a result, the use of wastewater for irrigation is forbidden or the wastewater must be highly treated and rigorously monitored in developed counties. Irrigation with sewage or sewage-contaminated surface waters in developing countries is fairly common and usually not regulated. While guidelines for wastewater reuse have been developed by the World Health Organization (WHO, 2006), their application in developing countries will remain difficult, due to inadequate institutional capability and general lack of financial resources.

The likelihood of the edible parts of the plants becoming contaminated during irrigation depends on a number of factors, including:

- growing location of the edible portion of the produce, e.g., distance from the soil or water surface
- frequency of irrigation
- surface of the edible portion, i.e., smooth, rough, or webbed
- type of irrigation method, i.e., furrow or flood irrigation, sprinkler, or drip

If the edible part of the crop grows in or near the soil surface, it is more likely to become contaminated than fruit growing in the aerial parts of the plant. In other cases, some produce surfaces are furrowed or have structures that retain water (e.g., a pepper versus a cantaloupe) and hence are more likely to be contaminated.

Table 10 Irrigated area (1,000 ha) by type of system, 1993 to 1997 (unless otherwise noted)[a]

Country	Surface	Sprinkler	Drip, spray, microirrigation
Chile	1,807.32	30.52	62.15
China	50,379.80	611.11	Not reported
India	49,330.00	700.00	71.00
Israel (1991)	Not reported	Not reported	104.30
Mexico	5,802.18	310.80	143.05
Spain (1988–1992)	Not reported	Not reported	160.00
USA (1988–1992)	14,002.14	9,226.84	606.00

[a]Source: FAO, 2007.

There are four types of irrigation systems: sprinkler, gravity-flow (furrow), and microirrigation systems (surface drip and subsurface drip irrigation). The type of irrigation system greatly influences the degree of crop contamination that occurs during irrigation. Stine et al. (2005a, 2005b) and others (Gerba and Choi, 2006) have compared coliphage contamination of cantaloupe, iceberg lettuce, and bell peppers by various methods of irrigation. Virus transfer to the lettuce was 4.2%, 0.02%, and 0.00039% for spray, furrow, and drip irrigation, respectively.

Area estimates of the type of irrigation systems used in different countries are inconsistently reported or are not up to date. The period from 1993 to 1997 is the last time that countries with important irrigation and food-export sectors reported their areas by irrigation type, as shown in Table 10 (FAO, 2007). For the majority of those listed countries, sprinkler irrigation is greater than drip, spray, or microirrigation. However, with the exception of the United States, the combined area of nonsurface irrigation (furrow, border, etc.) is less than one percent of total irrigated area, an observation supported by Namara et al. (2007) for India, and Postel et al. (2001) worldwide. Despite the small total area, drip irrigation is growing rapidly and is currently estimated to cover 2.8 million ha worldwide (Postel et al., 2001). Within the United States, California leads the way with a total of 730,500 ha (1.81 million acres) under drip and micro-irrigation (Burt et al., 2001).

Wastewater

Standards for irrigation water quality

Most standards for the microbial quality of irrigation water have been developed for the use of treated wastewater. Few standards have been suggested for water derived from other sources. Based on results of a study of irrigation waters in the western United States in the late 1960s, Geldreich and Bordner (1971) suggested an irrigation standard of 1,000 fecal coliforms per 100 ml based on the absence of *Salmonella* in irrigation waters, which had values below this level.

The World Health Organization recently revised its recommendations for the safe reuse of wastewater and greywater (WHO, 2006) by using a risk-based approach for development of treatment and microbial standards. The WHO guidelines for the safe use of wastewater in agriculture are based on a risk analysis approach and recommend treatment requirements for pathogen reduction, monitoring (verification of treatment performance), and management strategies. The level-of-protection goal is defined in terms of disability-adjusted life years (DALY), a measure which combines years of life lost by premature mortality with years lived with a disability, standardized or weighted by severity of illness (Pruss and Havelaar, 2001). The acceptable level of risk from consumption of pathogens on food is defined as 10^{-6} DALY (WHO, 2006). Based on this analysis, the minimum requirements for irrigation water for use on root crops are equal or less than 1,000 *E. coli* per 100 ml and zero helminth eggs per liter. This guideline is based on a wastewater treatment process that provides a 4-log (99.99%) reduction in pathogens (approximately equivalent to an *E. coli* of 1,000/100 ml in unchlorinated effluents), a 2-log pathogen reduction due to die-off between the last irrigation and consumption, and a 1-log reduction by washing of the salad crops or vegetables with water prior to consumption. The WHO believes this option provides the needed 7-log pathogen reduction for crops eaten uncooked. For options totally dependent on the treatment to remove pathogens (again a 7-log reduction) to the required level of acceptable risk, the *E. coli* level in irrigation water for crops eaten uncooked should be equal to or less than one *E. coli* per 100 ml.

Unfortunately, many assumptions on the ratio of *E. coli* to pathogens were made in the risk analysis because of the lack of data on pathogens in wastewater in developing countries. In addition, the reliability of the treatment has to be considered in the assessment of risks. Even treatment plants designed to produce high-quality reclaimed wastewater in developed countries like the United States do not produce pathogen-free effluents at all times, because of the occurrence of suboptimal operation of various processes in the treatment train. Even short-term (hours), suboptimal operation can result in the occurrence of pathogens in the treated wastewater. In a study of the quality of reclaimed wastewater in California, Tanaka et al. (1998) found that the processes in use were not reliable enough to recommend the treated wastewater for unrestricted irrigation of food crops to be eaten raw. Finally, some wastewater treatment processes may be very effective at removing *E. coli*, but very inefficient in removing other groups of pathogens. For example, chlorine disinfection is very good at reducing the levels of *E. coli* but has little effect on *Cryptosporidium* oocysts when chlorination is applied at conventional levels. Management of risks from wastewater irrigation must take all of these factors into consideration when developing guidance for wastewater used in agriculture.

Use of wastewater irrigation in the developing world

How common wastewater irrigation is in the developing world is not known with certainty, but it has been estimated that 20 million ha of crops are irrigated with raw or diluted sewage (Scott et al., 2004). The World Health Organization estimates that 10% of the world's population consumes food produced by irrigation with untreated wastewater (WHO, 2006). The percentage is considerably higher among populations in low-income countries with arid and semiarid climates. Irrigation of sewage-contaminated surface waters is common, although it may not occur as end-of-pipe wastewater for crop production. Ideally, restrictions are in place for the treatment of sewage before reuse and irrigation is limited to crops that pose the least risk to consumers. Van der Hoek (2004) gathered available data on the occurrence of wastewater reuse across the globe. The bulk of the untreated sewage from Mexico City is used for non-food crop irrigation. This accounts for half of the known 500,000 ha irrigated with wastewater in Latin America. Significant wastewater irrigation of food crops near the major cities occurs in Peru and Bolivia. Wastewater is directly used for irrigation in almost all towns in Pakistan, where vegetables are the most commonly irrigated crops, because they bring the highest prices. It has been estimated that at least 32,500 ha of land are irrigated directly with wastewater (Ensink et al., 2004). In China 20 billion cubic meters (5,280 billion gallons) of municipal wastewater was discharged in 1998 (Wei et al., 2000) with only 5% of it being treated. In Africa, Cornish and Kielen (2004) compared the use of untreated wastewaters in Ghana, where wastewaters are an important source of reliable irrigation water. In this study, microbial quality of irrigation water varied greatly (in terms of fecal coliforms) depending on location, dilution, and effects of natural remediation (e.g., antimicrobial effects of sunlight). Wastewater in these areas provides an important source of financial gain for many growers operating small plots across a wide area, pointing to the real challenges in regulating wastewater irrigation.

Compost and Night Soil

Night soil refers to human feces collected in buckets, pit latrines, or composting systems without a collection system. In developing countries, especially in rural areas, where collection systems are not available, this is the common method of waste disposal/treatment. Night soil is still commonly used in parts of eastern Asia, as it has been used for centuries as a fertilizer. In Asia, especially in China and Vietnam, the use of human excreta for crop production is still widespread. In one region of central Vietnam, more than 75% of the farmers studied used fresh or partly composted human excreta to fertilize their farms or gardens (Jensen et al., 2005). This practice likely accounts for the high prevalence of intestinal parasites in Vietnam. In some farming communities in north and central Vietnam, hookworm rates with human populations are 70% or more (Verle et al., 2003). It was estimated in 1993 that China produced 109 million tons of night soil per day by 200 million people in 450 cities lacking any treatment facilities (Bo et al., 1993). The untreated night soil is transported to rural areas where it is used in agriculture. Night soil generated in the rural areas has also been commonly used for garden crops. In areas where untreated night soil is used, the incidence of intestinal parasites has been as high as 93% compared with as low as 28% in areas where treated (stored) night soil was used.

Often night soil is stored in vaults or composted before use on agriculture land; however, prolonged storage and composting (10 months or longer may be required to kill intestinal parasites) are essential to reduce the risk from enteric pathogens (Feachem et al., 1983; Jensen et al., 2005). Vietnam allows the use of composted hygienic safe human excreta as a fertilizer in horticulture but has only recently defined the minimum required composting times (Jensen et al., 2005).

Shellfish-Growing Waters
Shellfish consumed raw have long been known to be associated with the transmission of infectious disease because these filter feeders have the ability to concentrate microorganisms, including pathogens, from water. In its quest for food, an oyster may filter as much as 1,500 liters of water per day (Gerba and Goyal, 1978). The use of strict standards in the United States for shellfish-harvesting waters and processing has virtually eliminated outbreaks from enteric pathogens (FDA, 2005), although occasional outbreaks from norovirus still occur from shellfish harvested in the United States (Berg et al., 2000). Enteric viruses in shellfish are the most common enteric pathogen of concern in shellfish and have been responsible for many outbreaks (Gerba and Goyal, 1978; Richards, 2006). Detection of enteric viruses in shellfish from around the world indicates that virus in shellfish is a common occurrence (Bosch et al., 2001; Kingsley et al., 2002; Beuret et al., 2003). Although the protozoan parasite *Cryptosporidium* has been detected in shellfish, no outbreaks have been documented (Gómez-Couso et al., 2003). Guidance is provided for bacteria standards for shellfish-growing waters and shellfish meat (FDA, 2005); however, the longer survival of enteric viruses in

marine waters and slower rate of depuration can limit their ability to assess the safety of the shellfish for consumption. That is why a sanitary survey of growing waters, harvesting, and processing is critical to reduce risks from enteric viruses.

According to the FDA (2005), approved shellfish-harvesting areas can average 14 fecal coliforms per 100 ml (using a five-tube most probable number method). These same standards apply to shellfish imported into the United States that are intended to be consumed raw. In contrast, the European Union regulates on the basis of fecal coliform levels in shellfish meats (Richards, 2006).

Outbreaks of disease associated with imported shellfish occur in many European countries (Bosch et al., 2001), Japan (Richards, 2006), and recently in the United States (Kingsley et al., 2002). Clams imported from China into the United States were associated with an outbreak of norovirus. Although import regulations require that clams from China be cooked, these clams were labeled as cooked but had the appearance of raw product (Kingsley et al., 2002).

Strict control of shellfish harvest waters is essential in any country where shellfish harvesting is practiced because of the ability of these organisms to concentrate pathogens. To ensure product safety, development of practical routine methods for virus detection and monitoring in growing waters and shellfish meats are needed.

Other Routes of Cross-Contamination

Fecally contaminated hands are another potential source of produce contamination during production, harvesting, and processing. Such contamination may result not only from ill individuals but also from the presence of asymptomatic excreters, which is likely to be greater in developing countries. Espinoza-Medina et al. (2006) found by using polymerase chain reaction (PCR) methodology evidence of *Salmonella* on the hands of 16.7% of the packers of cantaloupes in Mexico. More on this topic of hygienic deficiencies may be found in Chapter 5.

Use of any fecally contaminated water postharvest can result in contamination of produce. For example, Keraita and Drechsel (2004) found in Ghana that wash water was an important source of bacterial contamination of fresh produce sold in urban markets. The use of ice made from contaminated water may also be a factor (Cannon et al., 1991).

Use of human and animal feces and/or sewage for fish aquaculture is common in much of the world, especially in Asia, where much of the production takes place (Scholtissek and Naylor, 1988). Carp and tilapia are especially suited for growth in sewage/aquaculture systems. Fortunately, numerous studies suggest that while the gills, skin, and intestine of fish growing in these systems become contaminated with enteric bacteria and viruses, little contamination of the edible tissues occurs (Hejkal et al., 1983; Buras et al., 1985; Easa et al., 1995; Khalil and Hussein, 1997; Lan et al., 2007). As long as the harvested fish are handled and processed (e.g., cooked to proper temperatures) in a hygienic manner, there appears to be little risk to the consumer (Lan et al., 2007).

Use of municipal water in the developing world does not necessarily ensure that the water is safe for use in processing. Contamination of drinking water distribution systems is a major problem in the developing world, where an intermittent water supply is often the norm (Lee and Schwab, 2005). This results in loss of pressure and contamination of distribution systems by polluted water infiltrating the distribution system because of the negative pressure that occurs. This type of breakdown in the water supply system has been responsible for several drinking water-associated outbreaks in developing countries (Lee and Schwab, 2005). Other factors, such as poor infrastructure maintenance and lack of cross-connection control programs, exacerbate the problem.

THE WAY FORWARD

Multiple contamination sources and pathways pose risks to food safety, with solid and liquid waste handling and irrigation of produce and shellfish production representing major risk factors. A multiple-barrier approach is needed that identifies (and attempts to contain) the source of contamination, while at the same time it minimizes the transmission of contaminants to food, particularly through irrigation and postharvest handling. The WHO guidelines (WHO, 2006) for the safe use of wastewater include valid recommendations to manage these risks.

Clearly more information is needed on the microbiologic quality of water used in the production of food, especially for export so that the potential risks from contamination can be assessed and meaningful standards developed that protect human health. Fecal contamination of surface water will always be a problem. The use of animal manure and human solids wastes in the form of biosolids applied to soil are essential to the sustainability of agriculture and the recycling of nutrients. Management strategies are essential in the minimization of pathogen risks during agricultural production. These must reflect the types of crops, means of irrigation, harvesting, and processing in association with the level of environmental sanitation.

Finally, importers and exporters of food (that is potentially contaminated) and national authorities must establish mechanisms to share information regarding food contamination with potentially harmful agents as well as disseminate and provide resources for safe production and handling processes.

REFERENCES

Abbaszedegan, M., M. LeChevallier, and C. Gerba. 2003. Occurrence of viruses in U.S. groundwaters. *J. Am. Water Works Assoc.* **95**:107–120.

Berg, D. E., M. A. Kohn, T. A. Farley, and L. M. McFarland. 2000. Multi-state outbreaks of acute gastroenteritis traced to fecal-contaminated oysters harvested in Louisiana. *J. Infect. Dis.* **181**(Suppl 2):S381–S386.

Beuret, C., A. Baumgartner, and J. Schluep. 2003. Virus-contaminated oysters: a three-month monitoring of oysters imported to Switzerland. *Appl. Environ. Microbiol.* **69**:2292–2297.

Blumenthal, U. J., E. Cifuentes, S. Bennett, M. Quigley, and G. Ruiz-Palacios. 2001. The risk of enteric infections associated with wastewater reuse: the effect of season and degree of storage of wastewater. *Trans. R. Soc. Trop. Med. Hyg.* **95**:131–137.

Blumenthal, U. J., D. D. Mara, A. Peasey, G. Ruiz-Palacios, and R. Stott. 2000. Guidelines for the microbiological quality of treated wastewater used in agriculture: recommendations for revising WHO guidelines. *Bull. W. H. O.* **78**:1104–1116.

Bo, L., D. Ting-xin, L. Zhi-ping, M. Lou-wei, W. Zhu-xuen, and Y. An-xiu. 1993. Use of night soil in agriculture and fish farming. *World Health Forum* **14**:67–70.

Bosch, A., R. M. Pinto, and F. X. Abad. 2006. Survival and transport of enteric viruses in the environment, p. 151–187. *In* S. M. Goyal (ed.), *Viruses in Foods*. Springer, New York, N. Y.

Bosch, A., G. Sanchez, F. LeGuyader, H. Vanaclocha, L. Haugarreau, and L. M. Pinto. 2001. Human enteric viruses in Coquina clams associated with a large hepatitis outbreak. *Water Sci. Technol.* **43**:61–65.

Buras, N. 1976. Concentration of enteric viruses in wastewater and effluent—2 year survey. *Water Res.* **10**:295–298.

Buras, N., L. Duek, and S. Niv. 1985. Reactions of fish to microorganisms in wastewater. *Appl. Environ. Microbiol.* **50**:989–995.

Burt, C. M., D. J. Howes, and A. Mutziger. 2001. Evaporation estimates for irrigated agriculture in California. ITRC Paper no. P01-002, presented at the *2001 Irrigation Association Conference*, San Antonio, Texas, 4–6 November 2001.

Campo, N. C, I. L. Pepper, and C. P. Gerba. 2007. Assessment of *Salmonella typhimurium* growth in class A biosolids and soil/biosolids mixtures. *J. Residuals Sci. Technol.* **4**:83–88.

Cannon, R. O., J. R. R. B. Hirschhorn, D. C. Rodeheaver, P. R. Silverman, E. A. Brown, G. H. Talbot, S. E. Stine, S. S. Monroe, and D. T. Dennis. 1991. A multistate outbreak of Norwalk virus gastroenteritis associated with consumption of commercial ice. *J. Infect. Dis.* **164**:860–863.

Carr, R. M, U. J. Blumenthal, and D. D. Mara. 2004. Health guidelines for the use of wastewater in agriculture: developing realistic guidelines, p. 41–58. *In* C. A. Scott, N. I. Faruqui, and L. Raschid-Sally (ed.), *Wastewater Use in Irrigated Agriculture*. Commonwealth Agricultural Bureau International Publishing, Wallingford, United Kingdom.

Centers for Disease Control and Prevention (CDC). 2000. Foodborne outbreak of Group A rotavirus gastroenteritis among college students—District of Columbia, March–April 2000. *Morb. Mortal. Wkly. Rep.* **49:**1131–1133.

Chaidez, C., M. Soto, P. Gortares, and K. Mena. 2005. Occurrence of *Cryptosporidium* and *Giardia* in irrigation water and its impact on fresh produce industry. *Int. J. Environ. Health Res.* **15:**339–345.

Cooley, M. B., D. Chao, and R. E. Mandrell. 2006. *Escherichia coli* O157:H7 surival and growth on lettuce is altered by the presence of epiphytic bacteria. *J. Food Prot.* **69:**2329–2335.

Cornish, G. A., and N. C. Kielen. 2004. Wastewater irrigation – hazard or lifeline? Empirical results from Nairobi, Kenya and Kumasi, Ghana, p. 69–79. *In* C. A. Scott, N. I. Faruqui, and L. Raschid-Sally (ed.), *Wastewater Use in Irrigated Agriculture.* Commonwealth Agricultural Bureau International Publishing, Wallingford, United Kingdom.

Curriero, F. C., J. A. Patz, J. B. Rose, and S. Lele. 2001. The association between extreme precipitation and waterborne disease outbreaks in United States, 1948–1994. *Am. J. Public Health* **91:**1194–1199.

Cuthbert, J. A. 2001. Hepatitis A: old and new. Clin. *Microbiol. Rev.* **14:**38–58.

Easa, M. E. S., M. M. Shereif, A. I. Shaaba, and K. H. Mancy. 1995. Public health implications of waste water reuse for fish production. *Water Sci. Techol.* **32:**145–152.

Ensink, J., T. Mahmood, W. van der Hoek, L. Raschid-Sally, and F. P. Amerasinghe. 2004. A nation-wide assessment of wastewater use in Pakistan: an obscure activity or a vitally important one? *Water Policy* **6:**1–10.

Espinosza-Medina, I. E., F. J. Rodríguez-Leyva, I. Vargas-Arispuro, M. A. Islas-Osuna, E. Acedo-Felix, and M. A. Martinez-Tellez. 2006. PCR identification of *Salmonella*: potential contamination sources from production and postharvest handling of cantaloupes. *J. Food Prot.* **69:**1422–1425.

Food and Agriculture Organization (FAO). 2007. AQUASTAT Information System on Water and Agriculture. [Online.] http://www.fao.org/nr/water/aquastat/main/index.stm. Accessed 7 January 2008.

Fattal, B., Y. Lampert, and H. Shuval. 2004. A fresh look at microbial guidelines for wastewater irrigation in agriculture: a risk-assessment and cost-effectiveness approach, p. 59–68. *In* C. A. Scott, N. I. Faruqui, and L. Raschid-Sally (ed.), *Wastewater Use in Irrigated Agriculture.* Commonwealth Agricultural Bureau International Publishing, Wallingford, United Kingdom.

Food and Drug Administration (FDA). 2005. *Guide for the Control of Molluscan Shellfish.* National Shellfish Sanitation Program. Washington, DC.

Feachem, R. G., D. J. Bradley, H. Garelick, and D. D. Mara. 1983. *Sanitation and Disease. Health Aspects of Excreta and Wastewater Management.* John Wiley & Sons, New York, NY.

Gannon, V. P. J., T. A. Graham, S. Read, K. Ziebell, A. Muckle, J. Mori, J. Thomas, B. Selinger, I. Townshend, and J. Byrne. 2004a. Bacterial pathogens in rural water supplies in southern Alberta, Canada. *J. Toxicol. Environ. Health A* **67:**1643–1653.

Gannon, V. P. J., F. Humenik, M. Rice, J. L. Cicmance, J. E. Smith, and R. Carr. 2004b. Control of zoonotic pathogens in animal wastes, p. 409–425. *In* J. A. Cotruvo, A. Dufour, G. Rees, J. Bartram, R. Carr, D. O. Cliver, G. F. Craun, R. Fayer, and V. P. J. Gannon (ed.), *Waterborne Zoonoses.* IWA Publishing, London, United Kingdom.

Geldreich, E. E., and R. H. Bordner. 1971. Fecal contamination of fruits and vegetables during cultivation and processing for markets. *J. Milk Food. Tecnhol.* **34**:184–198.

Gerba, C. P. 2000. Assessment of enteric pathogens shedding by bathers during recreational activity and its impact on water quality. *Quant. Microbiol.* **2**:55–68.

Gerba, C. P., and C. Y. Choi. 2006. Role of irrigation water in crop contamination by viruses, p. 257–263. *In* S. M. Goyal (ed.). *Viruses in Foods.* Springer, New York, NY.

Gerba, C. P., and S. M. Goyal. 1978. Detection and occurrence of enteric viruses in shellfish: a review. *J. Food Prot.* **41**:743–754.

Gerba, C. P., and D. Kayed. 2003. Caliciviruses: a major cause of foodborne illness. *J. Food Sci.* **68**:1136–1142.

Gerba, C. P., J. B. Rose, C. N. Haas, and K. D. Crabtree. 1996. Waterborne rotavirus: a risk assessment. *Water Res.* **30**:2929–2940.

Gerba, C. P., and J. E. Smith. 2005. Sources of pathogenic microorganisms and their fate during land application of wastes. *J. Environ. Qual.* **34**:42–48.

Gómez-Couso, H., F. Freire-Santos, J. Martínez-Urtaza, O. Garcia-Martín, and M. E. Ares-Mazás. 2003. Contamination of bivalve molluscs by *Cryptosporidium* oocysts: the need for new quality control standards. *Intl. J. Food Microbiol.* **87**:97–105.

Guan, T. T., G. Blank, and R. A. Holley. 2005. Survival of pathogenic bacteria in pesticide solutions and on treated tomato plants. *J. Food Prot.* **68**:296–304.

Haug, R. 1993. *The Practical Handbook of Compost Engineering.* Lewis Publishers, Boca Raton, FL.

Hejkal, T. W., C. P. Gerba, S. Henderson, and M. Freeze. 1983. Bacteriological, virological and chemical evaluation of a wastewater-aquaculture system. *Water Res.* **12**:1749–1755.

Hejkal, T. W., E. M. Smith, and C. P. Gerba. 1984. Seasonal occurrence of rotavirus in sewage. *Appl. Environ. Microbiol.* **47**:588–590.

Ho, A. Y., A. S. Lopez, M. G. Eberhart, R. Levenson, B. S. da Silva, J. M. Roberts, P. A. Orlandi, C. C. Johnson, and B. L. Herwaldt. 2002. Outbreak of cyclosporiasis associated with imported raspberries, Philadelphia, Pennsylvania, *2000. Emerg. Infect. Dis.* **8**:783–788.

Hoang, L. M., M. Fyfe, C. Ong, J. Harb, S. Champagne, B. Dixon, and J. Issac-Renton. 2005. Outbreak of cyclosporiasis in British Columbia associated with imported basil. *Epidemiol. Infect.* **133**:23–27.

Huang, S. W. 2004. *Global Trade Patterns in Fruits and Vegetables.* Agriculture and Trade Report no. WRS-04-06. United States Department of Agriculture. Washington, DC.

Hurst, C. J., C. P. Gerba, and I. Cech. 1980. Effects of environmental variables and soil characteristics on virus survival in soil. *Appl. Environ. Microbiol.* **40**:1067–1079.

Hutchison, M. L., L. D. Walters, S. M. Avery, B. A. Synge, and A. Moore. 2004. Levels of zoonotic agents in British livestock manures. *Lett. Appl. Microbiol.* **39**:207–214.

Jensen, P. K., P. D. Phuc, A. Dalsgaard, and F. Konradsen. 2005. Successful sanitation promotion must recognize the use of latrine wastewater in agriculture—the example of Viet Nam. *Bull. W.H.O.* **83**:873–874.

Jimenez, B., C. Maya, E. Sanchez, A. Romero, L. Lira, and J. A. Barrios. 2002. Comparison of the quantity and quality the microbiological content of sludge in countries with low and high content of pathogens. *Water Sci. Technol.* **46**:17–24.

Kang, G., S. D. Kelkar, S. D. Chitambar, P. Ray, and T. Naik. 2005. Epidemiological profile of rotaviral infection in India: challenges for the 21st century. *J. Infect. Dis.* **192**(Suppl. S1):S120–S126.

Kasorndorkbua, C., T. Opriessing, F. F. Huang, D. K. Guenette, P. J. Thomas, X. J. Meng, and P. G. Halbur. 2005. Infectious swine hepatitis E virus is present in pig manure storage facilities on United States farms, but evidence of water contamination lacking. *Appl. Environ. Microbiol.* **71**:7831–7837.

Keraita, B. N., and P. Drechsel. 2004. Agricultural use of untreated urban wastewater in Ghana, p. 101–112. *In* C. A. Scott, N. I. Faruqui, and L. Raschid-Sally (ed.), *Wastewater Use in Irrigated Agriculture.* Commonwealth Agricultural Bureau International Publishing, Wallingford, United Kingdom.

Khalil, M. T., and H. A. Hussein. 1997. Use of waste water for aquaculture: an experimental field study at a sewage-treatment plant, Egypt. *Aquaculture Res.* **18**:859–865.

Kingsley, D. H., G. K. Meade, and G. P. Richards. 2002. Detection of both hepatitis A virus and Norwalk virus in imported clams associated with food-borne illness. *Appl. Environ. Microbiol.* **68**:3914–3918.

Lan, N. T. P., A. Dalsgaard, P. D. Cam, and D. Mara. 2007. Microbiological quality of fish grown in wastewater-fed and non-wastewater fishponds in Hanoi, Vietnam: influence of hygiene practices in local retail markets. *J. Water Health* **5**:209–218.

Lee, E. J., and K. J. Schwab. 2005. Deficiencies in drinking water distribution systems in developing countries. *J. Water Health* **3**:109–127.

Li, T. C., K. Chijiwa, N. Sera, T. Ishibashi, Y. Etoh, Y. Shinohara, Y. Murata, M. Ishida, S. Sakamoto, N. Takeda, and N. Miyamura. 2005. Hepatitis E virus transmission from wild boar meat. *Emerg. Infect. Dis.* **11**:1958–1960.

Maier, R., I. L. Pepper, and C. P. Gerba. 2000. *Environmental Microbiology.* Academic Press, New York, NY.

Mara, D. D. 2002. Waste stabilization ponds, p. 3330–3337. *In* G. Bitton (ed.), *Encyclopedia of Environmental Microbiology.* John Wiley & Sons, New York, NY.

Matthews, L., I. J. McKendrick, H. Ternent, G. J. Hunn, B. Synge, and M. E. Woolhouse. 2006. Super-shedding cattle and the transmission dynamics of *Escherichia coli* O157. *Epidemiol. Infect.* **134**:131–142.

Mattison, K., A. Shukla, A. Cook, F. Pollari, R. Friendship, D. Kelton, S. Bidwald, and J. M. Farber. 2007. Human noroviruses in swine and cattle. *Emerg. Infect. Dis.* **13**:1184–1188.

McKay, L. D., J. A. Cherry, R. C. Bales, M. T. Yahya, and C. P. Gerba. 1993. A field example of bacteriophage as tracers of fracture flow. *Environ. Sci. Technol.* **27**:1075–1079.

Mead, P. S., L. Slutsker, V. Dietz, L. F. McCaig, J. S. Bresee, C. Shapiro, P. M. Griffin, and R. V. Tauxe. 1999. Food-related illness and death in the United States. *Emerg. Infect. Dis.* **5**:607–625.

Medema, G. J., S. Shaw, M. Waite, M. Snozzzi, A. Morreau, and W, Grabow. 2003. Catchment characterization and source water quality, p. 111–158. *In Assessing Microbial Safety of Drinking Water, Improving Approaches and Methods.* Organization for Economic Co-operation and Development, Paris, France.

Meng, Q. S., and C. P. Gerba. 1996. Comparative inactivation of enteric adenovirus, polio virus, and coliphages by ultraviolet irradiation. *Water Res.* **30**:2665–2668.

Muniesa, M., J. Jofre, C. Garcia-Aljaro, and A. R. Blanch. 2006. Occurrence of *Escherichia coli* O157:H7 and other enterohemorrhagic *Escherichia coli* in the environment. *Environ. Sci. Technol.* **40**:7141–7149.

Namara, R., R. Nagar, and B. Upadhyay. 2007. Economics, adoption determinants, and impacts of micro-irrigation technologies: empirical results from India. *Irrig. Sci.* **25**:283–297.

Nyeleti, C., T. A. Cogan, and T. J. Humphrey. 2004. Effect of sunlight on the survival of *Salmonella* on surfaces. *J. Appl. Microbiol.* **97**:617–620.

Okafo, C. N., V. J. Umoh, and M. Galadima. 2003. Occurrence of pathogens on vegetables harvested from soils irrigated with contaminated streams. *Sci. Total Environ.* **311**:49–56.

Omisakin, F., M. MacRae, I. D. Orden, and N. J. Strachan. 2003. Concentration and prevalence of *Escherichia coli* O157 in cattle at slaughter. *Appl. Environ. Microbiol.* **69**:2444–2447.

Oragui, J. 2003. Viruses in feces, p. 473–476. *In* D. Mara and N. Horan (eds.), *The Handbook of Water and Wastewater Microbiology*. Academic Press, London, United Kingdom.

Postel, S., P. Polak, F. Gonzales, and J. Keller. 2001. Drip irrigation for small farmers: a new initiative to alleviate hunger and poverty. *Water Int.* **26**:3–13

Pruss, A., and A. Havelaar. 2001. The global burden of disease study and applications in water, sanitation and hygiene, p. 43–59. *In* L. Fewtrell and J. Bartram (ed.), *Water Quality: Guidelines, Standards and Health*. IWA Publishing, London, United Kingdom.

Reynolds, K. A., K. D. Mena, and C. P. Gerba. 2007. Risk of water borne disease in the United States. *Rev. Environ. Toxicol. Contam.* **192**:117–158.

Richards, G. P. 2006. Shellfish-associated viral disease outbreaks. p. 223–238. *In* S. M. Goyal (ed.), *Viruses in Foods*. Springer, New York, NY.

Robertson, L. J., and B. Gjerde. 2001. Occurrence of parasites on fruits and vegetables in Norway. J. Food. Protect. **64**:1793–1798.

Sanchez, G., A. Bosch, and R. M. Pinto. 2007. Hepatitis A virus detection in food: current and future prospects. *Lett. Appl. Microbiol.* **45**:1–5.

Scholtissek, C., and E. Naylor. 1988. Fish farming and influenza pandemics. *Nature* **331**:215.

Scott, C. A, N. I. Faruqui, and L. Rachid-Sally (ed.) 2004. *Wastewater Use in Irrigated Agriculture: Confronting the Livelihood and Environmental Realities*. Commonwealth Agricultural Bureau International Publishing, Wallingford, United Kingdom.

Seymour, I. J., and H. Appleton. 2001. Foodborne viruses and fresh produce. *J. Appl. Microbiol.* **91**:759–773.

Sivapalasingam, S., C. R. Friedman, L. Cohen, and R. V. Tauxe. 2004. Fresh produce: a growing cause of outbreaks of foodborne illness in the United States, 1973 through 1997. *J. Food Prot.* **67**:2342–2353.

Shannan, L. 1998. Irrigation development: proactive planning and interactive management. p. 251–276. *In* H. Bruins and L. Harvey (ed.), *The Arid Frontier*. Kluwer Academic Press, London, United Kingdom.

Smith, H.V., and A.M. Grimason. 2003. *Giardia* and *Cryptosporidium* in water and wastewater, p. 698–756. *In* D. Mara and N. Horan (ed.), *The Handbook of Water and Wastewater Microbiology*. Academic Press, London, United Kingdom.

Stine, S. W., I. Song, C. Y. Choi, and C. P. Gerba. 2005a. The effect of relative humidity on preharvest survival of bacterial and viral pathogens on the surface of cantaloupe, lettuce, and bell peppers. *J. Food Prot.* **68:**1352–1358.

Stine, S. W., I. Song, C. Y. Choi, and C. P. Gerba. 2005b. Application of microbial risk assessment to the development of standards for enteric pathogens in water used to irrigate fresh produce. *J. Food Prot.* **68:**913–918.

Straub, T. M., I. L. Pepper, and C. P. Gerba. 1993. Hazards from pathogenic microorganisms in land-disposed sewage sludge. *Rev. Environ. Contam. Toxcol.* **132:**55–91.

Sturdee, A. P., A. T. Bodley-Tickell, A. Archer, and R. M. Chalmers. 2003. Long-term study of *Cryptosporidium* prevalence on a lowland farm in the United Kingdom. *Vet. Parasitol.* **116:**97–113.

Tanaka, H., T. Asano, E. D. Schroeder, and G. Tchobanglous. 1998. Estimating the safety of wastewater reclamation and reuse using enteric virus monitoring data. *Water Environ. Res.* **70:**39–51.

Tierney, J. T., R. Sullivan, and E. P. Larkin. 1977. Persistence of poliovirus type 1 in soil and on vegetables grown in soil previously flooded with inoculated sewage sludge or effluent. *Appl. Environ. Microbiol.* **33:**109–113.

Thomas, K. M., D. F. Charron, D. Waltner-Toews, C. Schuster, A. R. Maarouf, and J. D. Holt. 2006. A role of high impact weather events in waterborne disease outbreaks in Canada, 1975–2001. *Int. J. Environ. Health Res.* **16:**167–180.

Thurston-Enriquez, J. A., P. Watt, S. C. Dowd, R. Enriquez, I. L. Pepper, and C. P. Gerba. 2002. Detection of protozoan parasites and microsporidia in irrigation waters used for crop production. *J. Food Prot.* **65:**378–382.

U.S. Environmental Protection Agency (USEPA). 1993. The standards for the use or disposal of sewage sludge. Final 40 CFR Part 503 rules. EPA 822/Z-93/001. USEPA, Washington, DC.

Van der Hoek, W. 2004. A framework for a global assessment of the extent of wastewater irrigation: the need for a common wastewater typology, p. 11–24. *In* C. A. Scott, N. I. Faruqui, and L. Raschid-Sally (ed.), *Wastewater Use in Irrigated Agriculture.* Commonwealth Agricultural Bureau International Publishing, Wallingford, United Kingdom.

Verle, P., A. Kong, N. V. De, N. Q. Thieu, K. Depraetere, H. T. Kim, and P. Dory. 2003. Prevalence of intestinal parasitic infections in northern Vietnam. *Trop. Med. Int. Health* **8:**961–964.

Wei, Y., Y. Fan, M. Wang, and J. Wang. 2000. Composting and compost application in China. *Resour. Conserv. Recycling* **30:**277–300.

World Health Organization (WHO). 2006. *Guidelines for the Safe Use of Wastewater, Excreta and Greywater,* vol. 2: *Wastewater Use in Agriculture.* WHO, Geneva, Switzerland.

World Bank. 2005. *Managing the Livestock Revolution.* Report no. 32725-GLB. World Bank, Washington, DC.

Zaleski, K. J., K. L. Josephson, C. P. Gerba, and I. L. Pepper. 2005. Survival, growth and re-growth of enteric indicator and pathogenic bacteria in biosolids, compost, soil, and land applied biosolids. *J. Residuals Sci. Technol.* **2:**49–63.

Imported Foods: Microbiological Issues and Challenges
Edited by Michael P. Doyle and Marilyn C. Erickson
© 2008 ASM Press, Washington, DC

Sanitation and Hygiene Deficiencies as Contributing Factors in Contamination of Imported Foods

5

Fengxia Dong and Helen H. Jensen

INTRODUCTION

Imports of agricultural products into the United States represent a significant share of many foods consumed in the United States today. Increased income, better transportation, and global food supply chains support vast trade in agricultural products. In 2006, the United States imported more than $64 billion worth of agricultural products. Beverages, tropical products, fruits, and vegetables led the import value (Dohlman and Gehlhar, 2007).

Developing countries often compete well in world agricultural commodity markets because their exported products can be produced at a lower cost or harvested in different seasons than in developed countries. As increased international food trade benefits consumers through lower prices and a greater quality and variety of food, it also introduces new food safety issues, revives previously controlled risks, and spreads contaminated food more widely across the globe. Although countries often have practices in place to regulate the products entering their borders, they have limited ability to regulate production outside of their borders. Because of the large amount of product traded and from many sources and variety, the sampling and testing required to guarantee safety of many shipments means that the importing country cannot guarantee all shipments of food as safe. Therefore, consumers in importing countries consume not only food from abroad, but also the services of other countries' food safety regimes (Mitchell, 2003).

As discussed in Chapter 1, changes in U.S. habits have included the consumption of more fresh fruit and vegetables that has necessitated increased

FENGXIA DONG, Center for Agricultural and Rural Development, Iowa State University, Ames, IA 50011-1070. HELEN H. JENSEN, Department of Economics, Center for Agricultural and Rural Development, Iowa State University, Ames, IA 50011-1070.

importation of the product. With such products, food sanitation and hygiene issues are more likely to be a concern as these foods are distributed, shipped, and consumed in fresh form. Consumption and imports of fish have also increased in the United States and in Europe. Imports of fish now account for >80% of total fish consumption in the United States compared with 55% in 1995 (Becker, 2007). Because fish are particularly prone to rapid pathogenic contamination resulting from unhygienic practices during and after fish are harvested, insufficient refrigeration, substandard processing, and poor packaging, the rise in fish consumption is also of concern.

SANITATION AND HYGIENE PROBLEMS AND HIGH COSTS OF FOOD-BORNE DISEASE

There are more than 200 diseases transmitted through food (World Health Organization [WHO], 2006). The agents causing food-borne disease include viruses, bacteria, parasites, toxins, pesticides, industrial chemicals, metals, and more recently, prions (WHO, 2006). The adverse health effects of food-borne diseases range from gastroenteritis (like abdominal pain, fever, vomiting, and diarrhea) to life-threatening conditions including cancer, birth defects, and neurological, hepatic, and renal syndromes. Many of the major food-borne diseases are caused by noroviruses, *Salmonella*, *Campylobacter*, *Shigella*, and *Escherichia coli*, and food-borne diarrheal disease is one of the most common illnesses worldwide (International Council of Nurses, 2001). Children, pregnant women, the elderly, and the immunocompromised are at greatest risk of both contracting food-borne diseases and suffering more serious adverse health effects. In developing countries, 1.8 million children under the age of five die each year because of diarrheal diseases (WHO, 2006).

The outbreaks of food-borne diseases have large impacts on consumers, producers, processors, and society, and lead to significant costs. A fundamental problem is that consumers usually cannot know whether food is safe until after consuming it. Food-borne illness imposes private costs on consumers, such as medical bills and lost days of work, and leads to consumers experiencing pain and suffering, and sometimes results in long-term health problems and even threats of death. Analysis of the economic impact of a *Staphylococcus aureus* outbreak in India showed that about 40% of the total cost of the outbreak was borne by the affected persons (WHO, 2006).

Countries as a whole incur substantial costs when food safety incidents occur. The U.S. Centers for Disease Control and Prevention (CDC) estimates that 76 million people suffer food-borne illnesses each year in the United States, accounting for 325,000 hospitalizations and more than 5,000 deaths. Health experts estimate that the yearly cost of all food-borne diseases in the

United States is \$5 to \$6 billion in direct medical expenses and lost productivity (Buzby et al., 1996). In addition, governmental agencies, such as the CDC and Food and Drug Administration (FDA), incur substantial costs in tracing the outbreak back to the contaminated product, investigating farm, processing, and packinghouse operations, inspecting more samples of imported products, and detaining imported products with potential risks to people's health.

Should a food safety incident occur, large quantities of the implicated food may need to be recalled and destroyed. The exported food may be rejected or condemned, and the food safety failure can result in significant economic losses of foreign exchange, which are very important to developing countries. Even those not directly associated with imported contaminated product, including domestic producers and other international suppliers, may suffer losses. When a food-borne illness is traced to a particular product instead of a particular grower, all suppliers of that food item may experience lost sales. Moreover, if a food-borne illness is traced to a particular product from a country, all firms in that country producing the product may face a import ban and endure costs. Such losses include the loss from interruptions of trade, lost market share, decreased prices, reduced capacity due to temporary or permanent plant closures, and costs and investment to meet higher and stricter requirements from importing countries. After the food-borne illness outbreaks due to *Cyclospora*-contaminated Guatemalan raspberries, U.S. imports of Guatemalan raspberries declined substantially. In 2001, U.S. imports of Guatemalan raspberries were 16% of 1996 levels (Calvin et al., 2003).

Understanding the inadequacy and problems of sanitation and hygiene deficiencies and their contribution to food contamination in exporting countries can lead to opportunities to reduce contamination in imported foods and to support efforts to reduce food safety problems in export partners. In this chapter, we focus on the sanitation and hygiene deficiencies in imported food. In general, those problems are more prominent in developing countries where food quality control is poorer than in industrialized countries. Sections in the chapter include presentation of evidence on food safety problems from imports, some cases related to hygiene and sanitation issues in exporting countries, and discussion of food safety activities in food production, especially in developing countries. Discussion about the challenges and progress suggest opportunities for developing safe food supply chains.

FOOD IMPORT REFUSALS TO THE UNITED STATES

The volume of FDA-regulated food imports has been rising rapidly. In FY2006, FDA received nearly 9 million shipments of imported food products, a level more than three times that of 10 years earlier. Despite this large volume, about 1% of the imported food shipments were physically examined, down from 1.7% in FY1996 when imports were less common (Becker, 2007). The relatively low rate of inspection reflects the significant increases in quantity of imports coupled with reduced resources for FDA food inspection. The quality and safety of imported foods are at risk because of the food safety practices of the exporting countries and the likelihood that sanitation and hygiene deficiencies would be imported along with the food items.

Because a large proportion of fresh fruit and vegetables and fish are imported from developing countries where poor internal food safety control systems, low food safety standards, and deficient food safety practices are prevalent, the trade in these products imposes high risk on U.S. consumers. For example, a 2003 FDA study found residue violations in 5.3% of imported fruits and 6.7% of imported vegetables sampled vs. 2.2% of domesticly produced fruits and 1.9% of domesticly produced vegetables. In total, violative residues were found in 6.1% of import samples versus 2.4% of domestic samples (Center for Food Safety and Applied Nutrition [CFSAN], 2005). And several years earlier, the FDA found *Salmonella* and *Shigella*, which can cause dysentery, in 4.4% of imported fruits and vegetables versus 1.1% of domestic products (CFSAN, 2001).

The FDA receives nearly 15 million import shipments of FDA-regulated products a year, which include human and animal drugs, medical devices and vitamins, in addition to food and food-related products. Vegetables and vegetable products and seafood products were the product categories with most frequent refusals (Becker, 2007). In a separate analysis, Table 1 lists the number of agricultural and aquatic products refused by the FDA in 2006 and the reasons for refusal by regions or countries for that year. Asian countries, excluding Japan, accounted for more than 43% of the total import refusals. Problems related to food additives, pesticide/veterinary drug residues, microbiological contamination (including bacteria, *Listeria*, and *Salmonella*), filth, and unapproved drug application accounted for more than 48% of the total for that region, with filth being the most prominent issue. Mexico is conspicuous in its large numbers of refusals as a single country, although it is the third largest source of agricultural imports to the United States, trailing only the European Union and Canada in terms of value. Food additives, pesticide/veterinary drug residues, filth, and labeling are the major problems found in imports from Mexico. Prominent violations for foods from Latin America were pesticide/veterinary drug residue regulations and standards.

For comparison, countries like Canada, Japan, and Australia have had fewer detentions. The major problems with European food products have mainly been related to labeling, drugs used in food production that are not

Table 1 FDA import refusals by regions or country in 2007[a]

Reason(s) for refusals	Africa	Latin America and Caribbean	Canada	Mexico	Europe	Asia (not incl. Japan)	Japan	Australia	Total
Food additives	25	108	84	259	237	812	22	14	1,561
Pesticide residues/veterinary drug	2	895	100	306	85	437	5	4	1,834
Microbiological contamination	17	91	18	127	87	586	34	5	965
Filth	25	217	54	684	140	1,098	41	4	2,263
Low-acid canned foods	29	100	5	15	198	355	63	26	791
Labeling	92	525	194	304	649	1,064	52	46	2,926
Other	2	33	25	49	92	452	22	7	682
Unapproved drug application	73	213	273	124	452	489	11	25	1,660
No process / not registered	111	234	121	71	1,224	1,767	119	31	3,678
Total	376	2,416	874	1,939	3,164	7,060	369	162	16,360

[a]Source: FDA, 2007.

approved in the United States, and incorrect filing of information. Africa does not have many detentions because its trade volume is relatively small.

Comparable evidence limited to food and food-related products by country for the period May 2006 to April 2007 indicates that Mexico (1,271 refusals) was the country with the highest number of food import refusals, followed by India (1,109 refusals) and China (720 refusals) (Becker, 2007). During this period, Canada led in dollar value of agricultural and/or seafood imports to the United States, followed by Mexico, then China. India ranked below the top 10 countries in terms of import volume.

Evidence from the FDA refusals, type, and country/region clearly reveals some common sanitation performance standard problems for developing countries. Hence, as developing countries become an increasingly important part of the global food system, their inadequate food safety practices are not solely their own problem. In these countries, the prominence of microbiological contamination of imported food items indicates that sanitation and hygiene deficiencies are a major contributing factor for contamination (Table 1).

SOURCES OF FOOD SAFETY HAZARDS IN IMPORTED FOOD PRODUCTS

From field to table, there is a long supply chain and many hazards can enter the food chain at different points. Both human and animal sources, largely through pathogen-contaminated feces, can, through interactions in the environment, lead to contamination of food and drinking water and ultimately transfer to humans (Prüss et al., 2002). Food-borne pathogens can be transmitted from human excreta though improper sanitation, water sources, insects, and soil. The transfer from food or drinking water to humans can be direct or through contaminated foods prepared for purchase or as processed. Farm production and food processing, with unhygienic facilities and hands or contaminated soil and water, can lead to disease being transmitted to people. While Chapter 4 provides information on the issue of animals as a contamination source, this chapter primarily focuses on human sources and exposures.

Lack of potable running water on production sites is a general problem in developing countries. In many exporting countries that supply raw food materials, standards of hygiene are poor. In agricultural production, untreated sewage/manure or sewage/manure-polluted wastewater for irrigating and fertilizing crops is prevalent (Horton, 1998). Poor sanitary practices can lead to contaminated water and vice versa. Potable water is critical for washing fruits and vegetables and icing seafood, and lack of it can result in a contamination with a host of bacteria, including *Salmonella* and *Shigella*.

Investment in potable water sources is a critical component of providing microbiologically safe water to production sites. Asia provides an example of the variation that can occur across countries in the same region. As large developing countries, India and China have a similar percentage of the population (approximately 25%) without adequate water. However, 85% of the population in India lack adequate sanitation services, whereas only 15% of Chinese face similar conditions. The other developing countries of Asia, on average, do worse than India and China in providing adequate potable water but provide sanitation services at a level between the Indian and Chinese extremes. Throughout Asia, about 40% of the people lack ready access to

The Case of Green Onions and Hepatitis A Outbreaks

Three outbreaks of hepatitis A occurred from September to November 2003 in Pennsylvania, Tennessee, and Georgia. More than 900 persons were affected and three died. FDA investigators pinpointed a link between the outbreaks and green onions from four firms in Mexico based on epidemiologic and traceback evidence. Hepatitis A is a liver disease caused by the hepatitis A virus that is transmitted by the fecal-oral route. Produce can become contaminated when a person who has a hepatitis A infection or whose hands are contaminated with hepatitis A virus has contact with the product or when the product is exposed to water contaminated with hepatitis A virus. After visiting the four Mexican firms and associated facilities, the FDA identified several factors that could have contributed to the spread of the infectious disease, including "poor sanitation, inadequate hand-washing facilities, questions about worker health and hygiene, the quality of water used in the fields, packing sheds, and the marking of ice."

At the time of the hepatitis A outbreaks in the United States, there were 26 green onion growers in Mexicali and San Luis Rio Colorado, the main production areas of summer production. Summer shipments (mid-May through early October) to the United States accounted for about 27% of total calendar year shipments in 2002. The output of the four firms named by the FDA as being the sources of the contaminated produce represented a small portion of the area's summer production of green onions and an even smaller share of total winter and summer production. But the problems of these four firms affected the entire industry.

The four firms, unlike many other firms, did not have third-party audits of good agricultural practices (GAPs) for their summer operations. Free-on-board (F.O.B.) prices of green onions were at $18.30 per box of medium green onions just before FDA's announcement implicating the product in the food-borne illness outbreaks on 15 November 2003. After the third FDA announcement identifying the four Mexican growers associated with the outbreak on 21 November, prices of green onions had declined to $12.43 per box. One week later, prices had declined further to $7.23 per box. For the 2-week period from 16 to 29 November 2003, estimated losses for Mexican growers totaled $10.5 million.

(Calvin et al., 2004)

potable water and about 60% lack adequate sanitation services (Horton, 1998).

Several outbreaks of hepatitis A associated with green onions imported from Mexico highlight the importance of sanitation and a clean water supply in the production and harvesting of fresh produce and how vulnerable the importing country can be to the sanitary conditions in the exporting country.

IMPORTANCE OF INFRASTRUCTURE, STANDARDS, AND INVESTMENT IN THE SUPPLY CHAIN

Fresh produce is a product particularly susceptible to water-borne contamination because it is often eaten raw. Its sourcing, and hence problems, may vary across the year. Exporting countries could benefit by taking strict standards or good practices as stimuli for investments in modernization of water sources and improved quality control practices in agriculture production and food processing.

A useful example is that of leading firms in Kenya's fresh produce industry and their efforts in the early 1990s to develop and maintain a supply network with British supermarkets. In order to provide consistent quality produce that met strict quality standards, the firms invested in sanitation and hygiene systems and other quality control programs to provide product to the premium-quality end of the market, including the growing demand for salads and other semi-prepared vegetable products. These firms, and their farmer suppliers, bore most of the costs of compliance and reaped most of the benefits. They constructed high-care processing facilities, invested in private laboratories, introduced improved lighting and water sanitation systems, advanced cold treatment and storage systems and facilities for worker hygiene, and developed full supply-chain traceability (World Bank, 2005).

As a result of the investment in improved facilities, water supply, and sanitation systems, as well as general compliance with the requirements of upscale supermarkets, the Kenyan exporters were able to see significant improvement of profit margin for goods processed and packaged for the high-end European market. Other benefits perceived by the exporters include regularity of demand, advance information from supermarket clients on market trends, certainty with respect to quality and hygiene specifications, and enhanced reputation. Kenya suppliers were able to increase the value of their fresh vegetable exports substantially, even during a period when EU imports from nonmember countries were flat. From 1991 to 2003, the value and volume of Kenya's exports of fresh vegetables increased fivefold (World Bank, 2005).

The opportunity to access and retain high-value markets often motivates firms to invest in sanitation controls and systems. However, success often requires significant investment, training, and control of the production and distribution process through the supply chain. The threat of import restrictions and bans on products are risks faced when there is loss of control and motivate improvements and investment in infrastructure and processing facilities. As illustrated in the boxed example, the fish industry in Kenya has been able to retain a market in Europe, but only after facing threats of losing the market due to lack of sanitation and hygiene control in the processing of fish products.

The Case of Kenya Fish Exports to the EU

Fish exports from Kenya have been a significant source of revenue and market for the Kenyan fish industry. The European Union was a very important market for Kenyan exports of fish, accounting for 59% of Kenyan exports of fish by volume and more than 95% of Kenyan exports of fresh fillets. Concerns about the safety of fish from Kenya first arose in November 1997 when Spain and Italy both banned fish imports from Kenya due to the presence of salmonellae. This ban caused a 34% decline of Kenya's fish exports to the EU. Two months later, following reports of a cholera outbreak in Kenya and neighboring countries in January 1998, the EU again banned imports of chilled fish products from Lake Victoria, citing poor hygiene standards. This time, Kenya's fish exports to the EU decreased 66% and foreign exchange earnings declined 32% from the previous year. A third ban in April 1999 followed a report that pesticides had been used in Lake Victoria to kill fish. This ban resulted in a further 68% decrease in fish exports and further loss of incomes, jobs, and foreign earnings.

In 1998 and 1999, exports of fresh fillets were about 86% less than in 1996. Following each of these cases, the EU imposed import restrictions, including testing and/or bans. The restrictions had major impacts on fish processors. Factories operated at or below capacity, and many companies that had the lowest hygienic standards and/or lacked the necessary processing facilities to switch to upgraded production of products had to close down. Moreover, the reduction of capacity and closure of fish-processing companies caused a loss of employment and negatively affected the fishing communities.

In order to reenter EU markets, the Kenyan government and processing sector invested heavily to satisfy the EU's hygiene requirements. Some regulations were introduced to ensure hygienic fish handling and processing. The Kenya Bureau of Standards published a code of hygiene practice for the handling, processing, and storage of fish, which applies to all fish regardless of whether for export or for the domestic market. This standard essentially harmonized Kenyan hygiene requirements for fish with those of the EU (Noor, 2002). In addition, community and stakeholders invested at landing sites and in processing facilities and training of personnel. As a result, the Kenyan fish-processing sector has become reliant on exports of fresh fillets to the EU, which provide a higher return than both local markets and exports of frozen fish to other developed countries.

(Henson et al., 2000)

FOOD SAFETY SYSTEM

From field to table, if some food safety defenses fail in the chain, food safety incidences can occur. Therefore, sanitary and hygiene-related risks are often not limited to one stage of production or processing. Interventions geared only to improving final product testing are of little value if the basic controls have not been effectively introduced earlier in the supply chain. In addition, it is very costly to test for their presence or if food safety incidents occur because of lack of sufficient food safety interventions throughout the food continuum. Hence, a preventative approach that pays attention to each point in the food chain by applying food safety interventions in food production and processing is preferred.

Sanitation and Hygiene Deficiencies in Food Handling and Processing Transmit Food Safety Hazards from People to Food

Transmission of sanitary and hygiene-related problems can occur through human excreta contamination of hands, water-borne sewage and nonrecycling latrines that contaminate soil, surface water and groundwater through direct contact or vector-borne by flies and insects (Prüss et al., 2002). Table 2 shows food safety activities in food production that can be applied during the food production process and emphasizes the fact that food safety issues may occur if there is failure at any point in the production, processing, and handling process. Applying good hygienic practices in facilities/establishments, hygienic handling of products by personnel, and clean water and transportation are major food safety activities in sanitation and hygiene control.

Production and processing establishments in developing countries often lack hygienic controls, various types of cleaning and sterilization equipment, and quality assurance management systems. In addition, traditional production practices and facilities may not comply with good agricultural practices (GAPs) required to ensure the safety of product. For example, the problems of insufficient and inappropriately placed toilets with hand-washing facilities, inadequate water supply for cleaning and food preparation, appropriate drainage, and regular removal of solid and liquid waste can lead to food safety issues at production and procession sites. The costs of modernization of basic infrastructure and facilities can be very high.

In addition to infrastructure, the lack of hygienic practices by personnel can introduce and transmit food safety problems. Many food safety problems are linked to the hygiene of the personnel who produce, process, and handle food (e.g., uncovered cuts or sores, touching food with unclean tools, coughing or sneezing over food, and not properly disposing of waste materials) (WHO, 2006).

Table 2 Food safety activities in food production[a]

Farm production	Transport of animals and agricultural products	Slaughter house, packing house, first distributor	Transport of products	Industrial process	Retailers, food services
Hygiene of facilities	Cleaning Disinfection	Hygiene of establishments	Cleaning vehicles	Hygiene of establishments	Hygiene of establishments
Hygiene of personnel		Hygiene of personnel	Cooling	Hygiene of personnel	Hygiene of personnel
Use of potable water		Ante- and post-mortem inspection and hygiene handling	Hygiene of personnel	Hygienic handling of products	Hygienic handling of products
Sewage contamination		Hygienic handling of products		Microbiological monitoring	Labeling
Control of use of agricultural pesticides		Monitoring of agrichemical residues		Labeling	
Control of use of veterinary pesticides, antibiotics, and hormones		Monitoring of residues of antibiotics, hormones, etc.			
		Microbiological monitoring			
		Labeling			

[a]Source: Walker, 1999; Unnevehr and Hirschhorn, 2001.

Deficiencies in Food Regulation and Systems of Enforcement Can Lead to Inadequate Food Safety Interventions in the Food System

As firms in countries with adequate sanitation and hygiene systems look to the added value in export markets, it is often the production environment and lack of control of standards in the food system that limit their ability to successfully compete. At the same time, buyers are becoming more interested in testing and assurances that the product is safe.

Sanitary and phytosanitary (SPS) problems have existed in agricultural production in China for many years but have only received worldwide attention since China's accession to the World Trade Organization (WTO). Causes for China's SPS problems can be attributed to its widely dispersed food system, lack of laboratory testing and food safety interventions, and problems in monitoring. These deficiencies are also common to developing countries.

The small scale of fresh produce and livestock operations in developing countries and the fact that they are relatively scattered across producing areas contribute to the abuse of agricultural chemicals and noncompliance with food safety regulations. For example, 92% of swine producers in China have an annual production with only one to five pigs (Ke, 2002). Controlling the use of chemicals and veterinary drugs in such a vast country—with more than 900 million farmers and countless household farming operations—is extremely difficult.

Poorly operating machinery, lack of assured refrigeration in distribution channels, and contamination of water supplies contribute to SPS problems. Small-scale farmers, food handlers, and processors have little or no motivation to comply with SPS regulations if they do not face penalties for noncompliance or if they face increased production risks. Even when large-scale, standardized production might develop, compliance with SPS standards can lead to significant increases in production costs and, in the short term, the potential loss of revenue can be a significant barrier to change. Fragmented food chains make it difficult to achieve traceable product flows and meet requirements for quality assurance.

Inefficient information systems and isolated markets mean that market information and other technical requirements are not communicated in an efficient manner. The lack of effective information channels across levels of government, industries, and regions means that even if some firms or industries confront SPS problems in export markets, other firms or industries likely would not be informed on a timely basis. In addition, most farmers do not have access to information about SPS standards, let alone to the resources required to comply with these standards such as appropriate technologies and scientific and technical expertise. Most producers have only a limited awareness of SPS measures in general and lack an understanding of their importance (Dong and Jensen, 2004).

CONSTRAINTS AND PROBLEMS RELATED TO HYGIENE AND SANITARY ISSUES IN DEVELOPING COUNTRIES

Compared with developed or high-income countries, in general, less developed countries are in the early stages of implementing food safety controls (Table 3). The low-income, developing countries rely more on targeted investments and training at key points in prevention because resource and capacity restrict them from having the more expanded control on food safety that more developed countries do. In developing countries, however, the responsibilities for food safety are undertaken mainly by one or two parties or not all. For example, consumers assume responsibility for food safety by mitigating the potential risks associated with food through their food choices and/or preparation methods (Henson, 2003). Even when government assumes the role and responsibility of food safety by regulation and oversight,

Table 3 Implementation of food safety controls based on level of economic development[a]

Activity	Evolution with level of economic development		
	Low income	Middle income	High income
Decision-making capacity	Stakeholder involvement in policy making Disease or hazard surveillance Participation in international standard-setting organizations		
	Qualitative risk assessment to inform risk management		Quantitative risk assessment by cost-benefit analysis
	Adopt international standards or standards of major import markers		Set standards according to local risk conditions or preferences
Provision of information	Targeted interventions for reduction of childhood illness and malnutrition	Consumers and industry education for improved food handling and preparation	Labeling and certification to inform consumers about production methods, product safety, and potential hazards
Prevention and control	Hygiene training at key points in food supply chain	Control of external or single-source hazards Phased implementation of standards Monitoring of key hazards in food supply chain	Control programs for single-source hazards Phased implementation of regulations for formal food sector Provide generic HACCP models for small processors and vendors
Infrastructure and research	Water supply Sanitation Marketing facilities Applied research to reduce hazards	Basic and applied research on many hazards	Sanitation and water supply Marketing infrastructure Research to develop hazard controls

[a]Source: Unnevehr and Hirschhorn, 2000.

it may be weak in some aspects. Moreover, the system lacks coordination between government, industry, and consumers. Such an environment creates major challenges for improvements in sanitation and hygiene systems that are inadequate for ensuring the safety of products for export markets.

As more formal food markets evolve, food producers and processors are expected to play an increasing role in providing safe foods. Furthermore, as incomes increase, consumers start to demand enhanced food safety controls through their market transactions and political process (Henson, 2003). In some cases, products destined for export markets might have higher food safety standards required by importing countries than those sold domestically. They require adequacy of water and raw materials, trained personnel, the absence of harmful pathogens, sanitation facilities, basic infrastructure, and the availability of technology critical to safe food production. Through adopting a high standard in sanitation and hygiene for exported food, suppliers can help maintain and improve market access and increase the firm's long-term competitiveness in the world market. However, such efforts are very costly and sometimes prohibitive for some countries, especially developing countries, which are poorly prepared to achieve the required control throughout the food system. In general, developing countries have poor regulatory structures, fragmented supply chains, less-trained personnel, low levels of sanitation, insufficient drinking water, and a lack of sufficient funds to invest in safety and quality control systems.

Institutional and Regulatory Systems

Failure to develop and implement national regulations that provide necessary food safety standards for agricultural and food production is a general problem in developing countries. Because regulations and standards are frequently outdated and inconsistent with or less restrictive than international commitments, it is difficult for producers in developing countries to produce products meeting the requirements of developed countries. In addition, the liability laws that enable parties who are harmed by contaminated food products to sue for damages are generally lax. Moreover, lack of coordination and unclearly delineated administrative responsibilities between government departments and agencies to control and ensure compliance make the regulations and standards less effective. The dispersed structure neither facilitates coordination nor supports effective implementation of food safety regulations. As a consequence, there often are in some areas many legal instruments dealing with food safety that overlap or repeat, but no controls at all in other areas. And sometimes some regulations have different requirements for the same food safety issues. Therefore, food safety control is not conducted in an efficient and effective manner.

An assessment of the food safety and agricultural health control systems of 33 countries in the Americas relative to the requirements to comply with and implement the SPS agreement was conducted by Bolanos et al. (2000) using three elements of key SPS capacity. The three elements include (i) institutional elements, defined as mechanisms through which national food safety and agricultural health interests are represented and defended, agreements implemented, and commitments acquired at the international level fulfilled; (ii) technological elements, defined as systems of food safety and agricultural health controls through which problems are identified, controls undertaken, and performance monitored; and (iii) regulatory elements, defined as systems of legislation relating to food safety and agricultural health issues and the mechanisms through which these are brought into compliance with international commitments. These investigators concluded that seven countries have food safety and agricultural health control systems that are generally favorable in their ability to meet the requirements of the SPS agreement, and 26 countries had SPS systems that are unfavorable. Among the surveyed countries, not surprisingly, the United States and Canada, followed by the major upper-middle-income economies of Latin America, were judged to have the greatest capacity. The low- and lower-middle-income countries, and particularly the island economies of the Caribbean, had much lower levels of capacity, particularly relating to technological and institutional elements. Moreover, across all countries, institutional capacity was the weakest element of the food safety and agricultural health control system.

Supply Chains

Large numbers of small-scale farms in developing countries contribute to the problem of food safety control. For example, in China, 70% of tea planting areas are operated by scattered households. Controlling the food safety practices in such scattered operations is extremely difficult. Small-scale producers have little or no motivation to comply with SPS regulations or use GAPs in the fields and good manufacturing practices (GMPs) in processing facilities and packinghouses if they do not face penalties for noncompliance or if they face increased production risks. Even when large-scale, standardized production might develop, compliance with SPS standards can lead to significant increases in production costs and, in the short term, the potential loss of revenue can be a significant barrier to change (Dong and Jensen, 2004).

Fragmented food supply chains for food products in developing countries worsen the noncompliance with regulations and difficulties of food safety control. Furthermore, traditional production methods may conflict with food safety requirements in export markets where production methods are very different (Henson, 2003). For example, EU hygiene requirements for fish and fishery products prohibit fish from making contact with wood, while most traditional fishing vessels in developing countries are constructed by wood (Henson, 2003).

Moreover, there is, in general, no traceback system established in developing countries. Food can be contaminated in each point of the supply chain, from planting/raising, management, harvest, transportation, storage, processing, and eventually the delivery of finished products to market. Food safety activities required in food production are listed in Table 2. From farm to the table, factors such as mishandled farm products, use of contaminated water, lack of hygiene in facilities and by personnel, improper cooling, inadequate cleaning of equipment, and infected persons touching foods can all contribute to disease outbreaks. A problem can occur in any step if food safety activities are not conducted appropriately and will potentially affect the safety of the final product.

Technology and Equipment

Main issues in developing countries regarding food safety controls are inadequate technology, equipment, and other infrastructure for testing product quality. Current inspection and testing technologies and instruments often do not provide the necessary measure or sensitivity for detecting possible SPS problems. In some cases, antiquated inspection methods and instruments are not able to detect the maximum residue levels set by developed countries for imported agricultural product, especially when more and more tolerances are set at very low doses (i.e., parts per million [mg/kg], parts per billion [μg/kg], and even parts per trillion [mμg/kg]) (Dong and Jensen, 2004). Out-of-date testing technology and laboratory facilities can result in questionable assurance results and limit the ability to conform to internationally accepted assessment procedures.

Production and processing establishments are often missing hygienic controls that should be implemented. Various types of cleaning and disinfection equipment and quality assurance management systems are also lacking. For example, on Lake Victoria of Kenya, fishing is still accomplished from wooden boats with a crew of two to four fishermen. Few of the boats are motorized, and the main technological advancement has been in the type and size of the nets. EU hygiene requirements for fish and fishery products prohibit fish from making contact with wood (Henson, 2003); hence, this antiquated production equipment, from the very beginning of production, makes Kenyan fish products in violation of EU hygiene requirements. In addition, facilities on landing beaches remain rudimentary and are often restricted to a covered area where fish are sold, and in some cases a landing jetty (Henson et al., 2000).

The problem of insufficient and inappropriately placed toilets with handwashing facilities is also common. There is no adequate water supply for cleaning and food preparation, appropriate drainage, and regular removal of solid and liquid waste; hence, food safety issues emerge at both production and processing sites. For example, there is rarely a source of potable running water, toilets, chilled storage facilities, or fencing to prevent entry of rodents and domesticated animals to the landing area on Kenya's Lake Victoria (Henson et al., 2000). Given such facilities, it is difficult to imagine that food safety and hygiene controls can be fully implemented. The cost of modernization of basic infrastructure and facilities is very high. For example, it is estimated that the cost of upgrading a single landing site on Lake Victoria to provide potable running water, cooling facilities, and some additional food safety upgrades has been estimated to be about $1.2 million (Henson et al., 2000). For many individual producers, this high cost would be prohibitive to upgrade necessary infrastructure and requires coordinated private efforts and public investment to meet the required changes.

Human Capital

Officials in government departments and agencies who perform food safety control functions are often poorly equipped or inadequately trained to fulfill their responsibilities. In addition, producers and processors often lack in knowledge about GAPs and GMPs. There are many food safety problems linked to the deficiencies in hygienic practices of personnel who produce or handle food (WHO, 2006). Although establishing basic hygiene and sanitation in food production, processing, and markets requires basic infrastructure related to sanitation, such as toilets and a safe water supply, it also requires instilling and encouraging hygienic practices in personnel and improved management of production processes, processing environment, and product flows to promote safer production and processing environments. It is necessary to communicate the principles of safe food production and handling practices and personal hygiene to achieve appropriate behavioral change in all participants in the food continuum.

CONCLUDING COMMENTS

With the constraints discussed above for developing countries, it is reasonable for suppliers to weigh the costs and benefits associated with participating in different market segments that have different food sanitation and hygienic requirements. In some cases, there may be large, profitable opportunities to service the domestic market, the regional market, or market segments in industrialized countries that have less stringent standards (World Bank, 2005). Alternatively, it may be more profitable to obtain a niche in more upscale markets by adopting higher hygienic and food safety standards.

A two-tiered system may be best for developing countries because improvement of an entire country's SPS situations may take time. There are some spillover effects from leading exporters with adoption of more stringent standards. Generally, leading exporters acquire or lease small-holder farms or encourage small holders to obtain certifications in order to obtain better control over their supplies. This backward integration may help small-holder farmers adopt necessary technical measures and reduce the harm placed by standards of compliance on the livelihood of the poor.

In some situations, where there is a large fixed cost to the infrastructure investment required by international markets, advances in sanitation control or training for the export market may bring along the domestic market at relatively little additional cost. In such cases, the benefit of investment in sanitation, hygiene, and other infrastructure that supports the safety of food in the system can have positive spillovers in the form of health benefits and safer food for domestic consumers.

As several of the examples provided above illustrate, improvements in sanitation and hygiene in developing economies are often driven by SPS requirements by developed countries in export markets and require coordination among participants in the food system. Since food safety problems can arise at many different points in the food supply chain, coordinated efforts among food producers, processors, regulators, and others involved in the food chain are essential for ensuring the safety of the food supply. On a political scale, this requires comprehensive food regulations and coordinated enforcement (WHO, 2006), whereas on the local scale, it needs the provision of appropriate training, infrastructure, and services. Foreign investment in infrastructure, facilities, and laboratory are often required, along with technical assistance from importing countries.

Because many of the problems of sanitation and hygiene require the efforts of individuals throughout the food production, processing, and distribution system, awareness of food safety matters needs to be raised for all who are involved. This includes farmers/fishermen, transporters, packers, processors, and distributors. The general hygiene requirements for food production and handling based on the recommendations contained in the Codex General Principles for Food Hygiene and GAPs/GMPs should be well known by all personnel who have contact with food in order to bring about changes in behavior to be consistent with food safety principles (WHO, 2006).

REFERENCES

Becker, G. S. 2007. U.S. food and agricultural imports: safeguards and selected issues. Congressional Research Service Report for Congress. Updated 8 November 2007. [Online.] http://www.nationalaglawcenter.org/assets/crs/RL34198.pdf. Accessed 3 December 2007.

Bolanos, E., P. Fernandez, E. Perez, and K. Walker. 2000. Situation of FTAA members for compliance with the WTO agreement on sanitary and phytosanitary measures. *Comunica* 4(15):40–42. [Online.] http://www.iica.int/comuniica/n_15/art.asp?art=9. Accessed 27 November 2007.

Buzby, J.C., T. Roberts, C.T. Lin, and J.M. MacDonald. 1996. Bacterial foodborne disease: medical costs and productivity losses. Economic Research Service, U.S. Department of Agriculture, Agricultural Economic Report no. 741. [Online.] http://www.ers.usda.gov/publications/aer741/aer741.pdf. Accessed 27 November 2007.

Calvin, L., L. Flores, and W. Foster. 2003. Case study: Guatemalan raspberries and *Cyclospora. In* L. J. Unnevehr (ed.), *Food Safety in Food Security and Food Trade*, International

Food Policy Research Institute 2020, Focus no. 10. [Online.] http://www.ifpri.org/2020/focus/focus10/focus10_07.pdf. Accessed 27 November 2007.

Calvin, L, B. Avendano, and R. Schwentesius. 2004 The economics of food safety: the case of green onions and hepatitis A outbreaks. Economic Research Service, U.S. Department of Agriculture, Outlook Report no. VGS30501. [Online.] http://www.ers.usda.gov/publications/vgs/nov04/VGS30501/VGS30501.pdf. Accessed 27 November 2007.

Center for Food Safety and Applied Nutrition, U.S. Food and Drug Administration (CFSAN). 2001. FDA survey of imported fresh produce. [Online.] http://www.cfsan.fda.gov/~dms/prodsur6.html. Accessed 27 November 2007.

Center for Food Safety and Applied Nutrition, U.S. Food and Drug Administration (CFSAN). 2003. FDA survey of domestic fresh produce. FY2000/2001 Field assignment. [Onlline.] http://www.cfsan.fda.gov/~dms/prodsu10.html. Accessed 27 November 2007.

Dohlman, E., and M. Gehlhar. 2007. U.S. trade growth: a new beginning or a repeat of the past? *Amber Waves*, September issue. Economic Research Service, U.S. Department of Agriculture. [Online.] http://www.ers.usda.gov/AmberWaves/September07/Features/USTrade.htm. Accessed 27 November 2007.

Dong, F., and H. H. Jensen. 2004. The challenge of conforming to sanitary and phytosanitary measures for China's agricultural exports. MATRIC Working Paper 04-MWP 8. [Online.] http://www.econ.iastate.edu/research/webpapers/paper_11475.pdf. Accessed 27 November 2007.

Food and Drug Administration (FDA). 2007. Import Refusal Report for OASIS. [Online.] http://www.fda.gov/ora/oasis/ora_oasis_ref.html. Accessed 27 November 2007.

Henson, S. 2003. The economics of food safety in developing countries. ESA Working Paper no. 03-19, Food and Agriculture Organization of the United Nations. [Online.] ftp://ftp.fao.org/docrep/fao/007/ae052e/ae052e00.pdf. Accessed 27 November 2007.

Henson, S., A.-M. Brouder, and W. Mitullah. 2000. Food safety requirements and food exports from developing countries: the case of fish exports from Kenya to the European Union. *Am. J. Agric. Econ.* **82:**1159–1169.

Horton, L. 1998. Food from developing countries: steps to improve compliance. *Food Drug Law J.* **53:**139–171.

International Council of Nurses. 2001. Food safety: an essential public health function of nurses. Fact Sheet (29/03/01). [Online.] http://www.icn.ch/matters_food.htm. Accessed 3 December 2007.

Ke, B. 2002. Perspectives and strategies for the livestock sector in China over the next three decades. Livestock Policy Discussion Paper no. 7. Food and Agriculture Organization, Livestock Information and Policy Branch, AGAL, August. [Online.] http://www.fao.org/ag/againfo/resources/en/publications/sector_discuss/PP_Nr7_Final.pdf. Accessed 27 November 2007.

Mitchell, L. 2003. Economic theory and conceptual relationships between food safety and international trade. *In International Trade and Food Safety: Economic Theory and Case Study.* Economic Research Service, U.S. Department of Agriculture, Agricultural Economic Report 828. [Online.] http://www.ers.usda.gov/publications/aer828/aer828d.pdf. Accessed 27 November 2007.

Noor, H. 2002. Sanitary and phytosanitary measures and their impact on Kenya, Standards and Trade workshop, Geneva, Switzerland. 16–17 May 2002. [Online.] http://www.unctad. org/trade_env/test1/meetings/standards/kenya3.doc. Accessed 4 January 2007.

Prüss, A., D. Kay, L. Fewtrell, and J. Bartram. 2002. Estimating the burden of disease from water, sanitation and hygiene at a global level. *Environ. Health Perspect.* **110**(5):537–542.

Unnevehr, L., and N. Hirschhorn. 2000. Food safety issues in the developing world. World Bank Technical Paper no. 469, World Bank, Washington, DC.

Unnevehr, L., and N. Hirschhorn. 2001. Designing effective food safety interventions in developing countries. [Online.] http://go.worldbank.org/Y0270X2TE0. Accessed 12 January 2007.

Walker, K. D. 1999. Political Dimensions of Food Safety, Trade, and Rural Growth. Presentation at the Inter-American Institute for Cooperation on Agriculture (IICA) World Bank Rural Week Conference, 26 March.

World Bank. 2005. The Impact of food safety and agricultural health standards on developing country exports. Summary of Report no. 31302, 10 January 2005. [Online.] http://www. standardsfacility.org/files/report31302.pdf. Accessed 27 November 2007.

World Health Organization (WHO). 2006. A guide to healthy food markets. [Online.] http://www.afro.who.int/des/fos/publications/healthy_food_book_2.pdf. Accessed 3 January 2007.

Imported Foods: Microbiological Issues and Challenges
Edited by Michael P. Doyle and Marilyn C. Erickson
© 2008 ASM Press, Washington, DC

Antimicrobial-Resistant Food-Borne Pathogens in Imported Food

6

Shaohua Zhao

INTRODUCTION

Antimicrobial resistance is an increasing public health threat impacting both human and animal health and is a cause for concern wherever antimicrobial agents are used in humans, animals, plants, and aquaculture. Although the benefits to human and animal health have been enormous over the past several decades with the introduction of antimicrobials, their use has also resulted in selecting for resistance among both pathogenic and commensal bacteria (National Research Council, 1998; Mead et al., 1999; Tollefson and Miller, 2000; McDermott et al., 2002; World Health Organization [WHO], 2002; Molbak, 2004; Serrano, 2005). As background to the relevance of antimicrobial-resistant food-borne pathogens in imported food, this chapter will first provide a brief explanation of the mechanisms by which antimicrobial resistance develops in microorganisms. The chapter will then focus on the prevalence of antimicrobial-resistant pathogens both in infections in humans and in food items, followed by a discussion on the Food and Drug Administration's (FDA's) surveillance of imported foods for this group of pathogens.

Use of antimicrobials in human medicine is a major factor contributing to development of antimicrobial resistance. Administration of therapeutic and subtherapeutic doses of antimicrobials to food animals, on the other hand, has also drawn great concern and debate for many years, due principally to the appearance and dissemination of multiple drug-resistant (MDR) pathogens that may be transmitted to humans through the food supply or by direct contact with animals (Mead et al., 1999; Fey et al., 2000; Phillips et al., 2004; White et al., 2004). Infection with antimicrobial-resistant bacteria has

SHAOHUA ZHAO, Division of Animal and Food Microbiology, Office of Research, Center for Veterinary Medicine, U.S. Food and Drug Administration, Laurel, MD 20708.

resulted in increased morbidity and mortality, blood stream infections, longer hospital stays, and more frequent readmissions for the hospital and intensive care unit (Helms et al., 2002 Martin et al., 2004; Varma et al., 2005b). Consequently, the cost of treating a resistant bacterial infection is much greater than a susceptible infection. The economic burden and the patient's suffering due to resistant bacterial infections are additional considerations.

While there is much disagreement on the health burden imposed by antimicrobial resistance of food-borne bacterial pathogens, it is generally agreed that antimicrobials, whether used for growth promotion, disease prevention or treatment, or pathogen control, can select for resistant bacterial pathogens, and that these pathogens can contaminate food originating from treated animals and result in human infections (White et al., 2004). The magnitude of the public health burden due to infections caused by antimicrobial-resistant food-borne pathogens differs by region and country. It can be affected by several factors, including human and veterinary antimicrobial use practices, process control at animal slaughter, food storage, and distribution systems, the availability of clean water, and proper cooking and home hygiene methods (WHO, 2002; White et al., 2004).

For the past 20 years, outbreaks associated with antimicrobial-resistant *Salmonella* and *Campylobacter* have been increasingly reported in many countries (Threlfall et al., 2000; Engberg et al., 2001; WHO, 2002), and resistance to fluoroquinolones and third-generation cephalosporins, which are important drugs to treat these bacterial infections, are of growing concern (Angulo et al., 2000; Fey et al., 2000; Zhao et al., 2001, 2003a). Moreover, global spread of these resistant clones has been occurring through international travel and imported food (Threlfall et al., 2000; WHO, 2002).

The rapid globalization of food production and trade has increased the potential for international incidents involving food contamination with microbial or chemical hazards. Indeed, imported foods have been recognized as an important source of *Salmonella*. Because antimicrobial resistance is spreading in food-borne pathogens, WHO recommends that the use of any antimicrobial agent for growth promotion in animals should be terminated if it is used in human therapeutics or known to select for cross-resistance to antimicrobials used in human medicine (WHO, 1997). The WHO also recommends monitoring antimicrobial resistance in important zoonotic food-borne bacteria and key indicator bacteria in food-producing animal populations and animal-based food products (WHO, 2002). This strategy enables the detection and development of risk management measures to reduce the transmission of resistant bacteria and resistance determinants from animals to humans, and the prudent use of antimicrobials in food animals and humans.

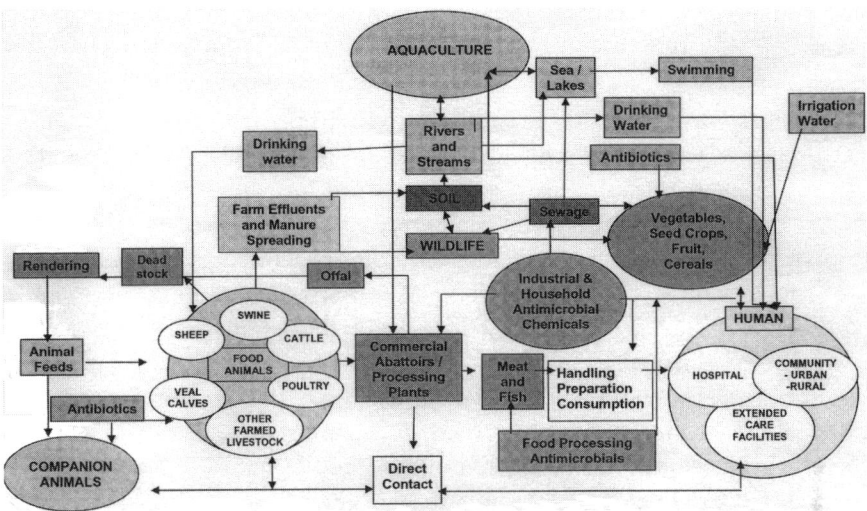

Figure 1 Dissemination of antimicrobial resistance. Possible pathways depicting transfer of antimicrobial resistance genes and/or antimicrobial-resistant bacteria (reproduced with permission from Rebecca Irwin [Prescoot et al., 2000]).

ANTIMICROBIAL RESISTANCE DEVELOPMENT AND DISSEMINATION

Antimicrobial resistance is an outcome of natural selection and an anticipated phenomenon of antimicrobial use, resulting from complex interactions among microorganisms, antimicrobial agents, and the environments in which they are brought together. In any large population of bacteria, some cells possess traits that enable them to survive in the presence of a toxic substance. Susceptible organisms are killed, whereas resistant populations thrive. With long-term antimicrobial use in a given environment, the number of resistant organisms can increase, causing a significant change in the microbial ecology. Additionally, resistant commensal and opportunistic bacteria can quickly become established as a component of the normal flora of various host species (McDermott et al., 2002; Witte, 2000b; Levy and Marshall, 2004). Moreover, the establishment of these resistant populations is cause for concern as transfer of resistance genes found on mobile genetic elements (plasmids, transposons, and integrons) to a susceptible recipient can occur via a single genetic event (transformation, conjugation, or transduction) while the organisms are present in animals, humans, or environments (Fig. 1) (Bush et al., 1995; Piddock, 1995; Witte, 2002a; Harbottle et al., 2006).

Antimicrobial uses in both agriculture and human medicine play a role in the development of resistance in some food-borne pathogens. The occurrence of resistance to some antimicrobials in nontyphoidal *Salmonella* has been associated with the use of antimicrobials in food animals. For example, *Salmonella enterica* serotype Typhimurium DT104 with the ACSSuT resistance profile (resistance to ampicillin, chloramphenicol, streptomycin, sulfamethoxazole, and tetracycline) was first isolated in the United Kingdom from sea gulls in the mid-1980s, but not from humans until 1989. During the next several years, the strain became prevalent in bovine, poultry, pig, and sheep. In another example, the increase of ciprofloxacin-resistant *Campylobacter jejuni* in humans was correlated with the approval of fluoroquinolone use in animals, particularly in poultry. In The Netherlands, a direct association between the licensing of fluoroquinolones for treating poultry infections and resistance developing in *Campylobacter* isolates from poultry was demonstrated along with a concurrent increase in resistance in human isolates. A similar scenario has been reported in the United States (Nachamkin et al., 2002), Spain (Wegener, 1999), and many other countries (Engberg et al., 2001). One of the most convincing reports illustrating the contribution of antimicrobial use as growth promoter in animals to the emergence of antimicrobial resistance was the use of nourseothricin in the former East Germany (Hummel et al., 1986). Farmers used this antimicrobial in swine feed from 1983 to 1990. *Enterobacteriaceae* isolated from both humans and animals in 1983 had little resistance to nourseothricin, but by 1985, transposon-encoded nourseothricin-resistant *Escherichia coli* were isolated from pigs and pork. By 1990, the resistant bacteria also spread to farmers, their families, and people living in the surrounding area where the pigs were grown. In addition, the resistance transposon spread to other bacterial species. Since nourseothricin was not used in humans, the resistance apparently resulted from feeding pigs nourseothricin as a growth promoter.

INFECTIONS ASSOCIATED WITH ANTIMICROBIAL-RESISTANT FOOD-BORNE PATHOGENS

In the United States, an estimated 76 million cases of food-borne illness, 325,000 hospitalizations, and 5,000 deaths occur each year (Mead et al., 1999). Over 90% of food-borne illnesses are caused by microorganisms, with *Campylobacter* and *Salmonella* being the leading causes of food-borne bacterial infections. Although data from the U.S. Centers for Disease Control and Prevention (CDC) revealed that the number of food-borne illnesses in the United States decreased in recent years (CDC, 2006b), the percentage of infections caused by many antimicrobial-resistant food-borne pathogens have increased over the past several years (CDC, 2006a). Moreover, some studies have revealed that antimicrobial-resistant pathogens caused more serious infections with increased mortality. For example, in an investigation of outbreaks caused by antimicrobial-resistant *Salmonella*, the CDC reported that the case-fatality rate was significantly greater for patients infected with antimicrobial-resistant *Salmonella* (4.2%) than for those with antimicrobial-susceptible infections (0.2%). More patients (22% of 13,286) in 10 outbreaks caused by resistant *Salmonella* were hospitalized than those (8% of 2,194 persons) infected in 22 outbreaks with pan-susceptible *Salmonella* strains (Varma et al., 2005a). In Denmark, patients infected with ACSSuT-resistant serotype Typhimurium strains were 4.8 times more likely to die, whereas quinolone resistance was associated with a mortality rate 10.3 times higher than the general Danish population (Helms et al., 2002).

A collaborative effort among the FDA, CDC, and the U.S. Department of Agriculture (USDA) was launched in 1996 to determine relationships between antimicrobial use in food animals and selected antimicrobial-resistant food-borne pathogens in humans. This program, known as the National Antimicrobial Resistance Monitoring System (NARMS), monitors antimicrobial resistance among food-borne enteric bacteria isolated from both clinical and food samples. The NARMS surveillance data are used to determine the potential role of antimicrobial-resistant pathogens in food-borne infections since no comprehensive information on such infections is available. A summary of the prevalence of antimicrobial-resistant food-borne pathogens in the United States over the past decades is presented in Table 1. To give a broader perspective of trends, the prevalence of antimicrobial-resistant pathogens in several other countries is also presented in Table 1.

In several cases, infections caused by antimicrobial-resistant *Campylobacter*, *Salmonella*, and other food-borne pathogens have been linked to foreign travel (Leegaard et al., 2000; Hakanen et al., 2001; Kassenborg et al., 2004). Countries to which persons with fluoroquinolone-resistant *Campylobacter* infections traveled included Spain, Portugal, and western Europe, in addition to Asia and Central and South America. The sources of travel-associated campylobacteriosis included undercooked poultry, contaminated water, raw milk, and cross-contaminated foods. The apparent association between the consumption of chicken and the acquisition of a ciprofloxacin-resistant *C. jejuni* infection among foreign travelers might point to the use of fluoroquinolones in veterinary medicine and animal husbandry. Many serovars of antimicrobial-resistant *Salmonella* were identified in travelers, with some serovars displaying resistance to multiple antimicrobials, including ampicillin, trimethoprim-sulfamethoxazole, chloramphenicol, tetracycline, and nalidixic acid. Nalidixic acid-resistant *Salmonella* were recovered from travelers from developed countries to India, Central America, Egypt, Morocco, Peru, and Kenya. Foreign travel and the consumption of imported foods

Table 1 Prevalence of antimicrobial-resistant enteric pathogens in the United States and other countries at selected time points between 1979 and 2006[a]

Organisms	Antimicrobial	Country	79–80	85–87	89–90	93	94–95	96	97	98	99	00	01	02	03	04	05	06
Non-Typhi *Salmonella*	Ampicillin	U.S.	9.0		13.0		15	20.7	18.3	16.5	15.6	15.9	17.4	12.9	13.6	12.0		
	Trimethoprim/sulfamethoxazole	U.S.	0.5		1.0		3.0	3.9	1.8	2.3	2.1	2.1	2.0	1.4	1.9	1.8		
	Ceftiofur	U.S.						0.2	0.5	0.8	2.0	3.2	4.1	4.3	5.5	3.4		
	Ceftriaxone	U.S.	0.0		0.0		0.0	0.0	0.1	0.0	0.3	0.0	0.0	0.2	0.4	0.6		
	Nalidixic acid	U.S.	0.0		0.0		0.5	0.4	0.9	1.4	1.0	2.5	2.6	1.8	2.3	2.6		
	ACSSuT[b]	U.S.	1.0		3.0		7.5	8.8	9.5	8.9	8.4	8.9	10.0	7.8	9.3	7.1		
Serotype Typhimurium	Ampicillin	U.S.						50.0	50.3	45.1	41.2	41.9	42.5	33.6	35.5	31.9		
	Amoxicillin/clavulanic acid	U.S.						2.6	3.4	4.5	2.8	6.3	6.2	7.6	5.2	4.7		
	Ceftiofur	U.S.						0.0	1.5	1.9	1.9	3.6	3.1	4.3	4.7	4.5		
	Ceftriaxone	U.S.						0.0	0.3	0.0	0.3	0.0	0.0	0.3	0.2	0.8		
	Nalidixic acid	U.S.						0.3	0.9	0.5	0.0	1.3	0.6	1.3	1.2	0.5		
	At least ACSSuT	U.S.	0.6					33.7	35.1	31.8	27.6	27.7	29.5	21.4	25.8	23.3		
	At least MDR-AmpC[c]	U.S.						0.0	1.2	1.1	0.6	2.0	1.2	1.8	2.2	2.6		
Serotype Newport	Ampicillin	U.S.						5.9	6.5	2.6	18.2	23.1	29.8	24.3	22.2	15.8		
	Amoxicillin/clavulanic acid	U.S.						2.0	0.0	2.6	18.2	22.3	26.6	22.2	21.3	15.3		
	Ceftiofur	U.S.						0.0	0.0	1.3	18.2	22.3	27.4	22.2	21.7	15.3		
	Ceftriaxone	U.S.						0.0	0.0	0.0	3.0	0.0	0.0	0.8	1.8	2.6		
	At least ACSSuT	U.S.						5.9	4.3	1.3	18.2	23.1	25.8	23.0	21.3	14.7		
	At least MDR-AmpC	U.S.						0.0	0.0	1.3	18.2	22.3	25.0	22.2	20.8	14.7		

Serotype	Antibiotic	Country	Resistance (%)
Serotype Enteritidis	Nalidixic acid	U.S.	0.9
Salmonella	Ampicillin	Spain	8, 44.0
	Tetracycline	Spain	1, 42
	Chloramphenicol	Spain	1.7, 26
	Nalidixic acid	Spain	0.1, 11.0
Salmonella	Nalidixic acid	Scotland	8.8, 1.4, 21.3
Serotype Typhimurium	Ampicillin	U.K.	1.0, 25
	Chloramphenicol	U.K.	1.5, 25
	Trimethoprim/sulfamethoxazole	U.K.	0.0, 25
Serotype Typhimurium	Tetracycline	Denmark	17.6, 24.8, 29.9, 26.7, 38.6, 41, 49, 46, 47, 53, 1.7
	Chloramphenicol	Denmark	7.6, 16.1, 24.1, 16.7, 14.6, 14, 13, 11, 23, 28, 2.0
	Ampicillin	Denmark	10.4, 22.5, 27.9, 22.7, 23.8, 34, 38, 40, 45, 56, 2.2
	Sulfonamide	Denmark	17, 27, 35, 29, 38, 44, 49, 40, 47, 53, 4.3
	Ciprofloxacin	Denmark	2.0, 5.8, 3.2, 2.2, 1.8, 1.2, 2.8, 2.6, 4.2, 4.7, 2.2
C. jejuni	Ciprofloxacin	U.S.	8.3, 0.0, 12.4, 13.8, 14.7, 18.4, 20.7, 17.2, 18.1, 2.2
	Nalidixic acid	U.S.	13.4, 15.5, 20.1, 16.0, 18.9, 21.3, 17.8, 18.4, 2.0
	Erythromycin	U.S.	1.4, 1.0, 0.7, 1.4, 1.0, 1.9, 1.2, 0.3, 0.3, 4.7
	Tetracycline	U.S.	47.8, 42, 46.1, 45.4, 39.2, 40.3, 41.3, 38.3, 46.9
C. jejuni	Nalidixic acid	Denmark	8.3, 11.9, 15.9, 25, 22, 17.2, 30.8, 28, 22.7, 3.9
	Erythromycin	Denmark	1.2, 0, 0, 2, 7, 0, 2.2, 4.7, 1.1, 4.7
	Tetracycline	Denmark	7.1, 7.3, 7.2, 22, 25, 15.1, 12.9, 24.3, 16, 13.6, 6.6
C. jejuni	Erythromycin	Canada	3, 2, 1, 12
	Tetracycline	Canada	43, 48, 68, 51
	Ciprofloxacin	Canada	10, 27, 26, 47

[a]From the following references: Threlfall et al, 1996; Prats et al, 2000; Gaudreau and Gilbert, 2003; Murray et al., 2005; CDC, 2006a; DANMAP, 2006.
[b] ACSSuT: resistance to ampicillin, chloramphenicol, streptomycin, sulfamethoxazole/sulfisoxazole, and tetracycline.
[c] MDR-AmpC: resistance to ACSSuT + amoxicillin-clavulanic acid and ceftiofur and decreased susceptibility to ceftriaxone (MIC $\geq$ 2 μg/ml).

contaminated with drug-resistant strains were suggested as causative factors in the increased incidence of antibiotic-resistant strains of *Salmonella enterica* serotype Enteritidis between 2000 and 2004, with resistance to nalidixic acid coupled with decreased susceptibility to ciprofloxacin associated with infections in England and Wales (Threlfall et al., 2006).

Epidemiologic studies have linked antimicrobial-resistant pathogen infections to the consumption of animal products. For example, the association between human MDR-AmpC *Salmonella enterica* serotype Newport infections and eating ground beef, drinking and eating unpasteurized dairy products, and living on a dairy farm, suggested that cattle may be an important reservoir for MDR-serotype Newport (Dunne et al., 2000; Zhao et al., 2003a; Berge et al., 2004), although horses, dogs, and pigeons can also carry the pathogen (Rankin et al., 2002). A more direct link was established in a U.S. study in which quinolone-resistant *C. jejuni* isolates from poultry were indistinguishable by DNA subtyping from clinical isolates (Smith et al., 1999). Similarly in an outbreak in France in 2003, MDR serotype Newport isolates containing the bla_{CMY} gene, encoding a cephalomycinase responsible for resistance to first-, second-, and third-generation cephalosporins, were identified in both clinical and imported horse meat samples (Espie et al., 2005). However, the country of origin of the imported meat could not be identified.

A recent study provided several examples of associations between clinical isolates and isolates obtained from imported food items (Aarestrup et al., 2007). Using pulsed-field gel electrophoresis (PFGE) molecular typing to differentiate isolates, one PFGE type of *Salmonella enterica* serotype Schwarzengrund that was frequently resistant to ciprofloxacin was found in clinical isolates from individuals in Denmark, Thailand, and the United States, in chicken meat from Thailand, and in dehydrated whole chili imported from Thailand to the United States. A second PFGE type that was nalidixic acid-resistant was found in clinical isolates from individuals in Denmark and the United States, and in catfish imported from Thailand to the United States.

Infections with fluoroquinolone-resistant *Campylobacter* are documented in many countries (Engberg et al., 2001), including Austria, Denmark, Finland, France, The Netherlands, Italy, Spain, Thailand, and the United Kingdom. The highest rates of quinolone-resistant *Campylobacter* infections in humans occurred in Thailand, with >90% followed by >70% in Spain, >38% in Austria, >35% in Italy, >30% in Finland, >30% in Denmark, >25% in France, >20% in The Netherlands, 18% in the United States, and 15% in the United Kingdom (Engberg et al., 2001). Increased fluoroquinolone-resistant *Campylobacter* infections often coincided with or followed the approval of fluoroquinolones in animal husbandry. The development of resistance to macrolides and fluoroquinolones has a significant impact on public health because these antimicrobials are advocated as first- and second-line drugs for the treatment of severe cases of *Campylobacter* infection. Because relatively little fresh chicken meat is currently imported into the United States, it is suggested that fluoroquinolone-resistant *Campylobacter* infections in this country result from consumption of domestic poultry products or foreign travel (Angulo et al., 2004). Concern with imported chicken and fluoroquinolone-resistant *Campylobacter*, however, may arise if imports of this commodity increase in the future.

E. coli O157:H7 isolates usually have lower rates of resistance to antimicrobials compared with other food-borne pathogens. The most common profile is resistance to tetracycline, streptomycin, and sulfamethoxazole; however, resistance to beta-lactamase antimicrobials such as ampicillin, amoxicillin-clavulanic acid, cephalothin, ceftiofur and ceftoxitin has also developed (CDC, 2006a). In any event, the significance of the antibiotic resistance of *E. coli* O157:H7 is debatable because antibiotic treatment of illness caused by this pathogen is contraindicated.

Concerns regarding the antimicrobial resistance of *Listeria monocytogenes* remain low as isolates from clinical, food, or environmental sources have little resistance to antimicrobials, with the exception of tetracycline (Poros-Gluchowski and Markiewicz, 2003; Shen et al., 2006; Zhang et al., 2007). Moreover, in a survey of 84 clinical isolates of *L. monocytogenes* collected between 1955 and 1997, the rate of resistance to penicillin, ampicillin, erythromycin, tetracycline, and chloramphenicol did not increase. Despite this observation, continued evaluation of antimicrobial resistance profiles of *L. monocytogenes* isolates should be continued in light of the demonstration that this pathogen can acquire antimicrobial resistance genes from foreign sources through movable genetic elements (Biavasco et al., 1996; Roberts et al., 1996; Charpentier and Courvalin, 1999).

ANTIMICROBIAL-RESISTANT *SALMONELLA* FROM IMPORTED FOODS

The FDA has regulatory oversight of about 80% of the U.S. food supply that includes inspection of both domestic and imported fresh produce, seafood, spices, dairy products, exotic meats, eggs, and ready-to-eat foods. Meat and poultry, on the other hand, are inspected by USDA. Although FDA has monitored imported foods for microbial contamination for many years, antimicrobial susceptibility testing was not performed on these isolates until 2000 (Zhao et al., 2003b; Koones, 2006). Hence, there are relatively few reports on the prevalence of antimicrobial-resistant food-borne pathogens in imported food. Currently, *Salmonella* is the only pathogen routinely tested for antimicrobial susceptibility. Because of limited data availability, this section will focus on antimicrobial-resistant *Salmonella* isolated from FDA-regulated imported products, in particular, seafood and fresh produce.

As discussed in Chapter 1, it is estimated that 86% of seafood, 17% of fresh vegetables, and 40% of fresh fruits consumed in the United States in 2005 were imported. Currently, 1% of imported food products under the jurisdiction of FDA are sampled and tested as part of the FDA inspection program. The FDA places greater emphasis for inspection on products having a history of the greatest occurrence of *Salmonella* contamination, such as shellfish. As a result, from 1999 to 2005, more than 30,000 food samples were tested by FDA, and the vast majority of those samples were imported foods. The overall *Salmonella* contamination rates of imported foods were 4 to 6%, with seafood having the highest contamination rate (Kiessling et al., 2005, 2006). Common serotypes included serotype Weltevreden, serotype Newport, serotype Lexington, serotype Saintpaul, serotype Thompson, serotype Typhimurium, and serotype Enteritidis. Examples of types of antimicrobial-resistant *Salmonella* isolates from 2000 to 2005 are listed in Table 2 (Zhao et al., 2003b, 2006; Kiessling et al., 2005, 2006).

In 2000, among 187 *Salmonella* isolates representing 82 serotypes recovered from imported foods tested for antimicrobial susceptibility, 15 (8%) isolates were resistant to at least one antimicrobial, and 5 (2.7%) were resistant to three or more antimicrobials. One isolate from Cambodian frozen anchovies exhibited resistance to six antibiotics. Twelve of the 15 resistant isolates were recovered from seafood, and three were from fresh produce (coriander and parsley) and cheese. Resistance to tetracycline (5%), was most often observed, followed by sulfonamides (4%) and streptomycin (3%). Antimicrobial resistance was not associated with specific *Salmonella* serotypes, as the 15 antimicrobial-resistant isolates represented 14 different serotypes. Nalidixic acid resistance (MICs from 128 to $\geq$ 256 µg/ml) was detected in four isolates recovered from either tilapia or catfish from Taiwan or Thailand. These resistant isolates also had decreased susceptibility to ciprofloxacin (Zhao et al., 2003b).

In 2001, among 208 *Salmonella* isolates, representing 66 serotypes recovered from imported foods, 23 (11%) isolates were resistant to at least one antimicrobial, and seven (3.4%) were resistant to three or more antimicrobials. Resistance to tetracycline (9%) was most often observed, followed by sulfamethoxazole (5%), streptomycin (4%), nalidixic acid (3%), and trimethoprim/sulfamethoxazole (2%). One serotype Schwarzengrund isolate recovered from squid imported from Taiwan was resistant to eight antimicrobials (Zhao et al., 2006).

Table 2 Multidrug-resistant *Salmonella* strains recovered from FDA-regulated imported foods from 2000 to 2005[a,b]

Year	Serotype	Resistance[c]	Product	Country
2000	Hadar	Nal-Str-Tet	Frozen catfish	Thailand
	Typhimurium	Smx-Str-Tet	Cheese	Denmark
	London	Smx-Str-Tet	Frozen shrimp	Vietnam
	Blockley	Kan-Nal-Str-Tet	Frozen catfish	Thailand
	Derby	Amp-AMC-Str-Chl-Smx-Tet	Frozen anchovies	Cambodia
2001	Bahrenfeld	Smx-Str-Tet	Scallops	Vietnam
	Schwarzengrund	Cot-Nal-Smx-Tet	Frozen catfish	Thailand
	Infantis	Cot-Kan-Nal-Smx-Tet	Chili power	Pakistan
	Bareilly	Amp-Chl-Cot-Smx-Tet	Frozen dace fish	Vietnam
	Ohio	Amp-Cot-Kan-Smx-Str-Tet	Paprika	Spain
	Schwarzengrund	Amp-Cip-Chl-Nal-Smx-Str-Tet	Dehydrated chili	Thailand
	Schwarzengrund	Cot-Chl-Gen-Kan-Nal-Smx-Str-Tet	Squid roll	Taiwan
2002	Enteritidis	Gen-Smx-Str	Frozen catfish	Thailand
	Typhimurium	Nal-Smx-Str	Guinea pig	Ecuador
	Montevideo	Smx-Str-Tet	Coriander	Bulgaria
	Typhimurium	Amp-Smx-Str	Dried oregano	Mexico
	Typhimurium	Amp-Str-Tet	Duck yolk	Vietnam
	Blockley	Nal-Smx-Str-Tet	Basil	Egypt
	Hadar	Nal-Smx-Str-Tet	Frozen catfish	Thailand
	Typhimurium	Amp-Smx-Cot-Tet	Frozen shrimp	Ecuador
	Newport	Amp-Chl-Smx-Str-Tet	Dried cuttle fish	Korea
	Group C2	Amp-Gen-NAl-Smx-Str-Tet	Basil	Egypt
	Haifa	Amp-Nal-Smx-Cot-Tet-Tri	Basil	Egypt
	Indiana	Amp-Chl-Nal-Smx-Str-Cot-Tet-Tri	Fish	China
	Kentucky	Amp-Cip-Gen-NAl-Smx-Str-Tet-Tri	Okra	Egypt
	Schwarzengrund	Amp-Chl-Gen-Nal-Smx-Cot-Str-Tet-Tri	Fish	Taiwan
2003	Typhimurium	Amp-Str-Tet	Duck yolk	Vietnam
	Typhimurium	Amp-Smx-Str	Oregano	Mexico
	Montevideo	Smx-Str-Tet	Coriander	Bulgaria
	Blockley	Nal-Smx-Str-Tet	Basil	Egypt
	Hadar	Nal-Smx-Str-Tet	Frozen catfish	Thailand
	Enteritidis	Nal-Smx-Str-Tet	Frozen squid	China
	Albany	Smx-Cot-Tet-Tri	Turmeric	Pakistan
	Stanley	Nal-Smx-Str-Tet	Instant seasoning	Pakistan
	Hadar	Amp-Gen-Nal-Smx-Tet	Frozen shrimp	China
	Haifa	Amp-Nal-Smx-Cot-Tet-Tri	Basil	Egypt
	Schwarzengrund	Amp-Chl-Gen-Nal-Smx-Cot-Str-Tet-Tri	Fish	Taiwan

(continued)

Table 2 Multidrug-resistant *Salmonella* strains recovered from FDA-regulated imported foods from 2000 to 2005[a,b] *(continued)*

Year	Serotype	Resistance[c]	Product	Country
2004	Agona	Str-Smx-Tet	Black pepper	Malaysia
	Agona	Str-Smx-Tet	Coconut milk	Philippines
	Albany	Smx-Tet-Cot	Turmeric	Pakistan
	Braenderup	Str-Smx-Cot	Breaded crab claws	Thailand
	Java	Nal-Str-Tet	Crabmeat	Venezuela
	Kentucky	Str-Smx-Tet	Frozen snapper	Vietnam
	Typhimurium	Str-Tet-Fis	Shredded coconut	Vietnam
	Weltervreden	Str-Smx-Tet	Shrimp	China
	Stanley	Nal-Str-Smx-Tet	Instant seasoning	Pakistan
	Thompson	Kan-Smx-Tet-Cot	Frog legs	China
	Typhimurium	Str-Smx-Tet-Cot	Shrimp	Philippines
	Weltevreden	Amp-Cep-Str-Tet	Milk fish	Philippines
	Havana	Chl-Str-Smx-Tet-Cot	Frozen lobster tail	Mexico
	Monophasic	Fox-Cep-Str-Smx-Tet	Cumin seeds	Turkey
	Virchow	Chl-Nal-Tet-Cot-Fis	Frog legs	China
	Heidelberg	Amp-Kan-Nal-Str-Tet-Fis	Fresh claw meat	Venezuela
	Enteritidis	Amp-Cep-Chl-Gen-Kan-Nal-Str-Smx-Tet	Imitation scallop	China
2005	Agona	Str-Tet-Fis	Frozen swaipangasius	Thailand
	Blockley	Kan-Nal-Tet	Basil	Egypt
	Corvalis	Str-Tet-Fis	Frozen yellow catfish	Malaysia
	Kentucky	Chl-Tet-Fis	Shelled pecans	Mexico
	Senftenberg	Str-Tet-Fis	Yellow fin tuna	Thailand
	Senftenberg	Chl-Gen-Kan	Tahini	Lebanon
	Typhimurium	Str-Tet-Fis	Frozen octopus	Philippines
	Weltevreden	Amc-Amp-Fox-Chl	Unicorn fish	Fiji
	Schwarzengrund	Amp-Cip-Gen	Frozen sharptooth broadhead fish	Thailand
	Typhimurium	Chl-Gen-Str-Tet-Cot	Fresh cilantro	Mexico
	Enteritidis	Amp-Chl-Nal-Str-Tet-Fis	Confectionary sunflower	China
	Typhimurium	Amc-Amp-Chl-Str-Tet-Fis	Frozen breaded shrimp	China
	Lansing	Amp-Chl-Gen-Kan-Nal-Str-Tet-Cot-Fis	Bass	Taiwan

[a]From the following references: Zhao et al., 2003b, 2006; Kiessling et al., 2005, 2006. Two antimicrobial susceptibility testing methods were used to generate above data; micro broth dilution was used for 2000, 2001, 2004, and 2005 isolates and disk diffusion was used for 2002 and 2003 isolates.

[b]Multidrug resistance is defined as resistance to ≥3 antimicrobials.

[c]Amc, Amoxicillin/clavulanic acid; Amp, ampicillin; Cip, ciprofloxacin; Cot, trimethoprim/sulfamethoxazole; Fis, Sulfissoxazole; Fox, Cefoxitin; Chl, chloramphenicol; Kan, Kanamycin; Gen, gentamicin; Nal, nalidixic acid; Smx, sulfamethoxazole; Str, streptomycin; Tet, tetracycline; Tri, trimethoprim.

Preliminary data reported by Kiessling et al. (2005) revealed 49 (14%) of 345 *Salmonella* isolates recovered from imported food in 2004 were resistant to at least one antimicrobial, and 15 (4.3%) were resistant to three or more antimicrobials. One serotype Heidelberg isolate from fresh claw meat from Venezuela was resistant to six antimicrobials, and a serotype Enteritidis isolate from imitation scallops from China was resistant to nine antimicrobials (Kiessling et al., 2005).

In 2005, 283 *Salmonella* isolates were recovered from FDA-regulated imported food, feed, and miscellaneous products (Kiessling et al., 2006). Forty-eight (17%) were resistant to one or more antimicrobial(s), and 17 (6%) were resistant to three and more antimicrobials. One serotype Lansing recovered from bass imported from Taiwan was resistant to nine antimicrobials. Major increases in resistance from 2004 to 2005 included ciprofloxacin from 0% to 4%, chloramphenicol from 4% to 12%, gentamicin from 1% to 9%, amoxicillin/clavulanic acid from 0% to 4%, and ampicillin from 3% to 12% (Kiessling et al., 2006). All four ciprofloxacin-resistant *Salmonella* isolates obtained in 2005 were from seafood from Southeast Asia (Vietnam and Thailand) and Lebanon. Seven of 11 chloramphenicol-resistant *Salmonella* were isolated from seafood from Argentina, China, Fiji, Indonesia, Lebanon, and Taiwan, and four were from confectionary sunflower, fresh cilantro, pecans, and chili powder imported from China, India, or Mexico. Five of seven gentamicin-resistant *Salmonella* isolates were from seafood from Bangladesh, Lebanon, Taiwan, Thailand, and Vietnam, and others from Mexican fresh cilantro and yucca powder. Two of four isolates with resistance to amoxicillin/clavulanic acid were from fish and shrimp from Fiji and China, respectively, and one each from Mexican pumpkin seeds and Egyptian dried mint leaves. Seven of 11 ampicillin-resistant *Salmonella* were recovered from seafood imported from Asian countries, whereas the remaining isolates were recovered from cheese, chili powder, confectionary sunflower and yucca powder imported from Colombia, Thailand, China, and Mexico, respectively. These data indicate that antimicrobial-resistant *Salmonella* in imported food is increasing with time, and resistant *Salmonella* is present in a wide variety of imported foods, especially aquatic products from Southeast Asia. Fresh produce and spices imported from many countries, particularly produce from Mexico, were also contaminated with antimicrobial-resistant *Salmonella*. The resistance profiles may reflect antimicrobials used in animal, plant, and aquaculture of the countries of origin.

Studies in Denmark revealed that *Salmonella* was more frequently isolated from imported meats than Danish meats, and the isolates from imported meats possessed higher rates of antimicrobial resistance (Skov et al., 2007), especially imported chicken and turkey meats (DANMAP, 2004, 2006). In particular, *Salmonella* isolates resistant to ampicillin, apramycin, gentamicin, nalidixic acid, spectinomycin, streptomycin, sulfamethoxazole, tetracycline, and trimethoprim were frequently isolated. In that same study, it was also reported that resistance to 12 antimicrobial agents (including nalidixic acid and ciprofloxacin) was significantly higher among serotype Typhimurium isolates from imported pork than pork from Danish pigs (DANMAP, 2004, 2006). In other documented cases, three isolates of serotype Virchow isolated from quail exported from France to Denmark in 2003 displayed resistance to ampicillin, ceftiofur, cephalothin, nalidixic acid, and tetracycline and reduced susceptibility to ciprofloxacin (Aarestrup et al., 2005). An ESBL-resistant serotype Heidelberg isolate from a boar exported from Canada to Denmark was also identified (Aarestrup et al., 2003). In addition, *C. jejuni* isolated between 1999 to 2003 from imported chicken had significantly higher resistance rates to tetracycline, chloramphenicol, nalidixic acid, and ciprofloxacin than *C. jejuni* from Danish-produced chicken meat (Andersen et al., 2006). It is evident from these data that antimicrobial-resistant food-borne pathogens can be transmitted internationally through global trade in food animals and their derived products.

In addition to imported foods for humans, FDA field laboratories assay both domestic or imported animal feed for microbial contamination. Following an outbreak of human salmonellosis associated with dog treats in Canada, 158 dog treats derived from pig ears and other animal parts were collected nationwide in 2000 by FDA and examined for the presence of *Salmonella*. Sixty-seven (43%) of the dog treats sampled were contaminated with *Salmonella*, including 8 (31%) domestic and 57 (43%) imported products (White et al., 2003). Fourteen samples were contaminated with more than one *Salmonella* serotype. Seventy-eight *Salmonella* isolates were tested for susceptibility to antimicrobial agents of human and/or veterinary importance. Resistance to tetracycline (26%), streptomycin (23%), sulfamethoxazole (19%), chloramphenicol (8%), and ampicillin (8%) was observed. Twenty-eight (36%) *Salmonella* isolates were resistant to at least one antimicrobial, whereas 10 (13%) isolates displayed resistance to four or more antimicrobials. Two isolates were identified as serotype Typhimurium DT104, with the characteristic ACSSuT R-type. Another serotype Typhimurium isolate was resistant to kanamycin in addition to the five antimicrobials above. One serotype Brandenburg isolate was resistant to eight antimicrobials.

ANTIMICROBIAL USE IN FOOD ANIMALS AND AQUACULTURE

It is difficult to describe practices of antimicrobial use in animal and aquaculture production in both developed and developing countries, because the level of control of antimicrobial usage is highly variable among countries

(WHO, 2002). With the exception of several European countries, there is little public information available regarding the total amounts of antimicrobials used in the production of various food animal species.

In Food Animal Production

Antimicrobials are used for several defined purposes in food animal production, including: (i) treatment of animals with overt signs of disease, (ii) prophylaxis in healthy animals at risk for contracting an infection, (iii) infection control (metaphylaxis) in a herd or flock having elevated morbidity or mortality, and (iv) improved feed efficiency and weight gain (growth promotion) in healthy animals. Examples of major classes of antimicrobials approved in the United States for use in food animals and their importance in human use are listed in Table 3.

Therapeutic use of antimicrobials is essential for maintaining animal health (McDermott et al., 2002). It is reported that more than 90% of the antimicrobials used in food animals are used for disease treatment or prevention (Tollefson and Miller, 2000; Phillips et al., 2004; Cabello, 2006). In modern production systems, food animals are generally raised in large groups. Given the close proximity of the animals to one another, and physiological and environmental stresses, underlying infections that may occur in a few animals can spread to others, including entire herds or flocks. Therefore, modern food animal production often requires aggressive approaches for disease control, which includes antimicrobial use. These medications are often administered through water and feed.

Continuous oral administration of low concentrations of antimicrobials may also increase feed conversion and weight gain in food animals. This practice can impose pressure for selection of antimicrobial resistance and accelerate the emergence of resistant bacteria that could infect humans through contaminated food or contact with animals.

The amount of antimicrobial agents used in non-human medicine in the European Union was estimated by the European Federation of Animal Health as 35% of the total usage in 1999. Non-animal agricultural applications are negligible in comparison to animal usage (Wassenaar, 2005). Antimicrobial usage in food animals in Denmark has been monitored and published annually in the DANMAP report since 1995, which is probably the most comprehensive data available to date. The total of prescribed antimicrobial agents used in food animals increased from 53,400 kg in 1990 to 115,150 kg in 2006; however, the total use of antimicrobials in animals in Denmark is considerably lower than that used before the termination of antimicrobial growth promoters in animal production in 1996 (DANMAP, 2006). The use of antimicrobials as growth promoters decreased from 79,308 kg in 1990 to

Table 3 Examples of antimicrobials approved for use in food animals in the United States and their corresponding importance in human medicine

Class	Antimicrobial agent	Mode of action/spectrum	Food-producing animal species	Examples of approved uses	Importance for human medicine (as class)[a]
Aminoglycosides	Gentamicin	Inhibit protein synthesis; broad spectrum	Turkeys, swine, chickens, cattle	Therapeutic use, disease prevention	Highly important
	Neomycin		Cattle, goats, sheep, swine, turkey	Therapeutic use	
	Streptomycin		Beef cattle, swine, chickens	Therapeutic use	
	Spectinomycin		Swine, chickens, turkeys, beef cattle	Therapeutic use, disease prevention	
Penicillins	Penicillin	Inhibit cell wall synthesis; narrow or broad spectrum dependent on individual agent	Swine, beef and dairy cattle, chickens, sheep, pheasants, turkeys, and/or quail	Therapeutic use, disease prevention, and/or growth promotion	Highly important
	Amoxicillin		Swine, beef cattle	Therapeutic use	
	Ampicillin		Beef cattle, swine	Therapeutic use	
Cephalosporin, first-generation	Cephapirin	Inhibit cell wall synthesis; narrow or broad spectrum dependent on individual agent	Dairy cattle	Intramammary for treatment of mastitis	Highly important
Cephalosporins, third-generation	Ceftiofur		Beef and dairy cattle, swine, sheep, goats, day-old broiler chicks, and turkeys	Therapeutic use	Critically important
Phenicol	Florfenicol	Inhibit protein synthesis; broad spectrum	Swine, cattle, catfish, and salmonids	Therapeutic use	Highly important
Lincosamides	Lincomycin	Inhibit protein synthesis; narrow spectrum	Swine, chickens	Therapeutic use, disease treatment/control, and/or growth promotion	Not ranked

Class	Drug	Mechanism	Animals	Use	Ranking[a]
Macrolides	Tylosin	Inhibit protein synthesis; narrow spectrum	Beef and dairy cattle, chickens, turkeys, swine, honey bees	Therapeutic use, disease prevention, and/or growth promotion	Critically important
	Tilmicosin		Beef and dairy cattle, sheep, swine	Therapeutic use	
	Erythromycin		Turkeys, cattle and dairy cattle, chickens, swine	Therapeutic use	
Polypeptides	Bacitracin	Inhibit cell wall synthesis; narrow spectrum	Turkeys, chickens, pheasants, beef cattle, quail, swine	Disease prevention, and/or growth promotions	Not ranked
Fluoroquinolones	Enrofloxacin	Inhibit DNA synthesis; broad spectrum	Beef cattle (excluding veal calves)	Therapeutic use	Critically important
	Danofloxacin				
Streptogramins	Virginiamycin	Inhibit protein synthesis; narrow spectrum	Swine, chickens, turkeys	Therapeutic use, disease prevention, and/or growth promotions	Highly important
Sulfonamides	Sulfadimethoxine	Inhibit folic acid synthesis; broad spectrum	Beef and dairy cattle, poultry, swine, catfish, salmonids	Therapeutic use	Highly important[b]
	Sulfamethazine		Swine, beef and dairy cattle, chickens, turkeys	Therapeutic, disease prevention, and/or growth promotions	Highly important[b]
Tetracyclines	Chlortetracycline, tetracycline, and oxytetracycline	Inhibit protein synthesis; broad spectrum	Beef cattle, dairy cows, poultry, sheep, swine, catfish, trout, salmon, lobsters	Therapeutic use, disease prevention, and/or growth promotion	Highly important

[a]Ranking is derived from FDA/CVM Guidance for Industry (http://www.fda.gov/cvm/Documents/fguide152.pdf).
[b]Potentiated sulfonamide compounds are ranked as critically important.

5 kg in 2004, with the use of most growth promoters terminated between 1999 and 2001. Since 2000, there has been no reported use of antimicrobial growth promoters for animals produced in Denmark (DANMAP, 2006). The antimicrobial resistance monitoring data revealed that resistance in different zoonotic bacteria has not uniformly been reduced since the ban. Many factors, such as mechanism of resistance, bacterial species or serotypes, animal origin, source of isolates (clinical cases versus the healthy animals), and prevalence of bacterial pathogens in the different animal environments, could influence the results. DANMAP 2006 reported that from 1999 to 2006, a significant increase in resistance to tetracycline ($P < 0.0001$), chloramphenicol ($P < 0.0001$), ampicillin ($P < 0.0001$), and sulfonamide ($P < 0.0001$) was observed among serotype Typhimurium isolated from swine, and from 2003 to 2006, streptomycin resistance ($P = 0.003$) also increased significantly (DANMAP, 2006). These increases coincided with increased consumption of tetracycline and sulfonamide/trimethoprim over the same period, whereas the consumption of macrolides penicillins with extended spectrum and aminoglycosides was slightly reduced. Although the overall consumption of antimicrobials in swine decreased marginally from 2004 to 2006, the proportion of fully susceptible isolates continued to decrease in 2005 and 2006. While the resistance patterns STR-SUL-TET and AMP-STR-SUL-TET continued to increase, this is most likely due to coresistance as the increase in consumption of some antimicrobials, such as tetracycline and sulfonamides, may select for these resistance patterns. Only a small number of serotype Typhimurium isolates was available from poultry and cattle, which made it difficult to determine differences in the occurrence of resistance from year to year. Erythromycin resistance in *Campylobacter coli* isolates from pigs decreased from 71% in 1997 to 13% in 2006, and resistance to tetracycline remained at a low level, although the consumption of tetracycline in swine increased from 1996 to 2006. Therefore, the increase in tetracycline use in pigs observed from 1999 to 2006 did not have a substantive effect on the occurrence of tetracycline resistance in *C. coli* isolates. *C. jejuni* isolated from broilers from 1996 to 2006 remained at a low level of resistance. Resistance in indicator bacteria such as enterococci to many antimicrobials is decreasing following the discontinuance of use of antibiotics as growth promoters. For example, there was decreased resistance to virginiamycin, avoparcin, and avilamycin by *Enterococcus faecium* isolated from broilers and broiler meat after the use of streptogramin, glycopeptides, and avilamycin was discontinued on poultry farms. *E. faecium* isolated from pigs and pork showed decreasing resistance similar to these antimicrobial agents following the ban.

There are in many developing countries no enforced policies regarding antimicrobial use in agriculture or associated drug residue tolerance in food.

Currently, few published data regarding antimicrobial use in food animals in developing countries are available. Several studies indicate that overuse of antimicrobials in food animal production is common in some developing countries (WHO, 2002). For example, large quantities of antimicrobials are routinely used during production of swine, poultry, and shrimp in Thailand for disease treatment and prevention (WHO, 2002). In some African countries, veterinary antimicrobials are easily accessible and under low levels of control by government authorities. The fact that expired antimicrobial drugs have been given new labels and sent to developing countries for use in animal production is another issue of concern. In addition, antimicrobials are used illegally or abused in some countries and, while the government has regulations for antimicrobial use, it lacks law enforcement. This situation is a serious problem in many developing countries in both human and veterinary medicine as well as in animal husbandry (WHO, 2002).

In Aquaculture

The use of antimicrobial agents in aquaculture is complicated and of great concern (Petersen et al., 2002; Cabello, 2006; Koones, 2006). Antimicrobials are usually administered directly into the water where the aquatic organisms live. Fish or shrimp are also given antimicrobials in their feed, and occasionally in baths or injections. Depending on the country, a wide variety of antimicrobials in large amounts, including nonbiodegradable antimicrobials, may be used and reach sediment at the bottom of the raising pens. From there, antimicrobials can leach from feed and feces and diffuse into the sediment, subsequently being washed by currents to distant sites. Once in the environment, these antimicrobials can be ingested by wild fish and other organisms, including shellfish. Residual antimicrobials can remain in the sediment at high levels for extended periods of time, exerting selective pressure, thereby altering the composition of the microflora of the sediment and selecting for antibiotic-resistant bacteria and affecting newly growing fish or shrimp (Khachatourians, 1998; Petersen et al., 2002; Holmstrom et al., 2003; Cabello, 2006). Studies on the occurrence of antimicrobial resistance in fish pathogens and environmental bacteria in four Danish rainbow trout farms revealed that high levels of individual and multi-antimicrobial resistance were found in *Flavobacterium* and *Aeromonas* species, indicating that fish farming had a major influence on the antimicrobial resistance of several groups of bacteria associated with aquacultural environments (Schmidt et al., 2000)

Many antimicrobials effective in human medicine are also used in aquaculture worldwide for treatment of bacterial infections in salmon, catfish, trout, and other commercially raised fish (Table 4) (Serrano, 2005; WHO,

Table 4 Major antimicrobial drugs (and classes) used in aquaculture in some locations in the world[a]

Antimicrobial agents (and class)	Administration	Importance in human medicine
Amoxicillin (aminopenicillins)	Oral	Critically important
Ampicillin (aminopenicillins)	Oral	Critically important
Chloramphenicol (amphenicols)	Oral/bath/injection	Important
Florfenicol (amphenicols)	Oral	Important
Erythromycin (macrolides)	Oral/bath/injection	Critically important
Streptomycin, neomycin (aminoglycosides)	Bath	Critically important
Furazolidone (nitrofurans)	Oral/bath	Important
Nitrofurantoin (nitrofurans)	Oral	Important
Oxolinic acid (quinolones)	Oral	Critically important
Entrofloxacin (fluoroquinolones)	Oral, both	Critically important
Flumequine (fluoroquinolones)	Oral/bath/injection	Critically important
Oxytetracycline, chlortetracyclines, tetracycline (tetracyclines)	Oral/bath/injection	Highly important
Sulfonamides (sulfonamides)	Oral	Important

[a]From report of a Joint FAO/OIE/WHO, Seoul, Korea, 2006 (WHO, 2006).

2006). The most frequent fish infections treated with antimicrobials are skin ulcers, diarrhea, and blood sepsis. The microorganisms responsible for these infections belong to bacterial families that can also cause infections in humans. Therefore, the transfer of antimicrobial resistance is highly probable. The use of antimicrobials in aquaculture in many countries is controlled by veterinarians. In the United States, oxytetracycline, sulfadimethoxine, ormethoprim, and florfenicol have been approved for use in controlling bacterial infections in salmon and catfish. In the United Kingdom, oxytetracycline, oxolinic acid, amoxicillin, and trimethoprim-sulfadiazine are approved for use in aquaculture. In Norway, benzylpenicillin+dihydrostreptomycin, florfenicol, flumequine, oxolinic acid, oxytetracycline, and co-trimazine are used in aquaculture. In Mexico, enrofloxacin and oxytetracycline are also used in aquaculture. In Denmark, the number of antimicrobials used in fish farms has been reduced greatly, with the total amount of antimicrobial consumption by farmed fish being reduced by approximately half from 4 tons in 2002 to 2.4 tons in 2004, which corresponds to a decrease from 105 to 70 mg of antimicrobial per kilogram of meat produced.

A wider range of antimicrobials are used in aquaculture in Asian countries than in the aforementioned countries (Holmstrom et al., 2003). For example, 74% of farms in Thailand used one or more types of antimicrobials prophylactically, some on a daily basis. At least 13 different antibiotics were used in shrimp ponds, including tetracycline, chlortetracycline, oxytetracycline, ciprofloxacin, enrofloxacin, norfloxacin, oxolinic acid, perfloxacin, sulfonamides, chloramphenicol, gentamycin, trimethoprim, tiamulin, plus 15 other unidentified antimicrobials. Many farmers were not well informed about efficient and safe application practices. Some incorrectly employed antimicrobials in pond management to prevent or treat viral diseases (Holmstrom et al., 2003). The widespread use of fluoroquinolones among the farms (e.g., norfloxacin and ciprofloxacin) is a particular cause for concern, considering their importance for treatment of a broad range of human pathogens.

Many studies have revealed that the bacterial flora in the environment surrounding aquaculture sites contain a large number of antimicrobial-resistant bacteria and these bacteria harbor new and previously uncharacterized resistance determinants of which this aquatic environment can select. These new resistance determinants have the potential of being transmitted through horizontal gene transfer to bacteria in the terrestrial environment, including human and animal pathogens (Cabello, 2006). In Southeast Asia, antimicrobial-resistant *Salmonella, Aeromonas hydrophila,* and *Aeromonas shigelloides* occurred in ponds in which antimicrobials were used routinely (Serrano, 2005). Resistance to tetracycline, oxytetracycline, furazolidone and sulfamethoxazole/trimethoprim was most common among these bacteria. In Thailand, *Salmonella* and *Vibrio cholerae* are commonly resistant to chloramphenicol, sulfamthoxazole/ trimethoprim, and gentamycin (Boonmar et al., 1998; Dalsgaard et al., 2000), and have been frequently isolated from tropical Asian shrimp farms (Reilly and Twiddy, 1992). In the Philippines, high rates of antimicrobial resistance were found among *Vibrio harveyi* and other bacteria isolated from shrimp ponds in which antimicrobials had been used. Resistance to oxytetracycline, furazolidone, oxolinic acid and chloramphenicol was most common, and multidrug resistance was widespread among many of the isolates. In the 1980s, when Taiwanese shrimp farming collapsed, one of the causative factors was the indiscriminate use of antibiotics, resulting in the development of resistant strains of shrimp pathogens (Lin, 1989).

The risk of development of antimicrobial-resistant pathogens from use of antimicrobials in aquaculture is a significant public health concern, especially in developing countries where there is limited control of antimicrobial use, and farmers may use indiscriminantly any antimicrobial they can obtain. Furthermore, other factors such as high organic loads, water temperature, pH and salinity in intensively managed tropical fish and shrimp ponds favorable for growth of microorganisms can influence the potential for contamination with antimicrobial-resistant bacteria (Reilly and Twiddy, 1992). In addition to microbial contamination, antimicrobial residues in seafood products produced on these farms are also of great concern.

CONCLUDING REMARKS

Imported foods, especially aquatic food products, can be contaminated with antimicrobial-resistant food-borne pathogens. The intercontinental transmission of antimicrobial-resistant pathogens via foods underscores the potential impact of agricultural and human antimicrobial use on consumer health worldwide. Global trade of foods is expected to increase in dramatic proportion in the future; hence, efforts to provide greater public health protection to foods should address antimicrobial-resistant food-borne pathogens in imported food products. It is essential that antimicrobials be appropriately used in humans and food animals, including aquaculture species, on a global basis to preserve the efficacy of existing drugs and to limit the risk of transfer of resistant food-borne pathogens to humans.

Global surveillance of food-borne pathogens and sharing of antimicrobial susceptibility data among developed and developing countries are also needed to monitor trends and emerging resistance phenotypes. Recently, the WHO launched the Global Salm Surv program (http://www.who.int/salmsurv/en/) in an attempt to strengthen and enhance the capacities of national and regional laboratories in the surveillance of *Salmonella* and *Campylobacter* infections. This initiative may be an important step in the quest to better control antimicrobial-resistant food-borne pathogens in the modern world.

ACKNOWLEDGMENTS
I thank Dr. David White for his insightful comments in the preparation of this chapter and Dr. Steve Yan for the information on antimicrobial agents approved to be used in food animals in the United States. I also thank Connie Kiessling for sharing data on antimicrobial resistance among *Salmonella* isolates from imported foods.

REFERENCES

Aarestrup, F. M., H. Hasman, and L. B. Jensen. 2005. Resistant *Salmonella virchow* in quail products. *Emerg. Infect. Dis.* 11:1984–1985.

Aarestrup, F. M., R. S. Hendriksen, J. Lockett, K. Gay, K. Teates, P. F. McDermott, D. G. White, H. Hasman, G. Oørensen, A. Bangtrakulnonth, S. Pronreongwong, C. Pulsrikarn, F. J. Angulo, and P. Gerner-Smidt. 2007. International spread of multidrug-resistant *Salmonella* Schwarzengrund in food products. *Emerg. Infect. Dis.* 13:726–730.

Aarestrup, F. M., M. Lertworapreecha, M. C. Evans, A. Bangtrakulnonth, T. Chalermchaikit, R. S. Hendriksen, and H. C. Wegener. 2003. Antimicrobial susceptibility and occurrence of resistance genes among *Salmonella enterica* serovar Weltevreden from different countries. *J. Antimicrob. Chemother.* 52:715–718.

Andersen, S. R., P. Saadbye, N. M. Shukri, H. Rosenquist, N. L. Nielsen, and J. Boel. 2006. Antimicrobial resistance among *Campylobacter jejuni* isolated from raw poultry meat at retail level in Denmark. *Int. J. Food Microbiol.* 107:250–255.

Angulo, F. J., K. R. Johnson, R. V. Tauxe, and M. L. Cohen. 2000. Origins and consequences of antimicrobial-resistant nontyphoidal *Salmonella*: implications for the use of fluoroquinolones in food animals. *Microb. Drug Resist.* **6**:77–83.

Angulo, F. J., J. A. Nunnery, and H. D. Bair. 2004. Antimicrobial resistance in zoonotic enteric pathogens. *Rev. Sci. Tech.* **23**:485–496.

Berge, A. C., J. M. Adaska, and W. M. Sischo. 2004. Use of antibiotic susceptibility patterns and pulsed-field gel electrophoresis to compare historic and contemporary isolates of multi-drug-resistant *Salmonella enterica* subsp. *enterica* serovar Newport. *Appl. Environ. Microbiol.* **70**:318–323.

Biavasco, F., E. Giovanetti, A. Miele, C. Vignaroli, B. Facinelli, and P. E. Varaldo. 1996. In vitro conjugative transfer of VanA vancomycin resistance between enterococci and listeriae of different species. *Eur. J. Clin. Microbiol. Infect. Dis.* **15**:50–59.

Boonmar, S., A. Bangtrakulnonth, S. Pornruangwong, S. Samosornsuk, K. Kaneko, and M. Ogawa. 1998. Significant increase in antibiotic resistance of *Salmonella* isolates from human beings and chicken meat in Thailand. *Vet. Microbiol.* **62**:73–80.

Bush, K., G. A. Jacoby, and A. A. Medeiros. 1995. A functional classification scheme for beta-lactamases and its correlation with molecular structure. *Antimicrob. Agents Chemother.* **39**:1211–1233.

Cabello, F. C. 2006. Heavy use of prophylactic antibiotics in aquaculture: a growing problem for human and animal health and for the environment. *Environ. Microbiol.* **8**:1137–1144.

Centers for Disease Control and Prevention (CDC). 2006a. National antimicrobial resistance monitoring system for enteric bacteria (NARMS): 2003 human isolates final report. U.S. Department of Health and Human Service, CDC, Atlanta, GA.

Centers for Disease Control and Prevention (CDC). 2006b. Preliminary FoodNet data on the incidence of infection with pathogens transmitted commonly through food—10 States, United States, 2005. *Morb. Mortal. Wkly. Rep.* **55**:392–395.

Charpentier, E., and P. Courvalin. 1999. Antibiotic resistance in *Listeria* spp. *Antimicrob. Agents Chemother.* **43**:2103–2108.

Dalsgaard, A., A. Forslund, O. Serichantalergs, and D. Sandvang. 2000. Distribution and content of class 1 integrons in different *Vibrio cholerae* O-serotype strains isolated in Thailand. *Antimicrob. Agents Chemother.* **44**:1315–1321.

DANMAP. 2004. *Use of Antimicrobial Agents and Occurrence of Antimicrobial Resistance in Bacteria from Food Animals, Foods, and Humans in Denmark.* Danish Institute for Food & Veterinary Research, Soborg, Denmark.

DANMAP. 2006. *Use of Antimicrobial Agents and Occurrence of Antimicrobial Resistance in Bacteria from Food Animals, Foods, and Humans in Denmark.* Danish Institute for Food & Veterinary Research, Soborg, Denmark.

Dunne, E. F., P. D. Fey, P. Kludt, R. Reporter, F. Mostashari, P. Shillam, J. Wicklund, C. Miller, B. Holland, K. Stamey, T. J. Barrett, J. K. Rasheed, F. C. Tenover, E. M. Ribot, and F. J. Angulo. 2000. Emergence of domestically acquired ceftriaxone-resistant *Salmonella* infections associated with AmpC beta-lactamase. *J. Am. Med. Assoc.* **284**:3151–3156.

Engberg, J., F. M. Aarestrup, D. E. Taylor, P. Gerner-Smidt, and I. Nachamkin. 2001. Quinolone and macrolide resistance in *Campylobacter jejuni* and *C. coli*: resistance mechanisms and trends in human isolates. *Emerg. Infect. Dis.* **7**:24–34.

Espie, E., H. De Valk, V. Vaillant, N. Quelquejeu, F. Le Querrec, and F. X. Weill. 2005. An outbreak of multidrug-resistant *Salmonella* enterica serotype Newport infections linked to the consumption of imported horse meat in France. *Epidemiol. Infect.* **133**:373–376.

Fey, P. D., T. J. Safranek, M. E. Rupp, E. F. Dunne, E. Ribot, P. C. Iwen, P. A. Bradford, F. J. Angulo, and S. H. Hinrichs. 2000. Ceftriaxone-resistant *Salmonella* infection acquired by a child from cattle. *N. Engl. J. Med.* **342**:1242–1249.

Gaudreau, C., and H. Gilbert. 2003. Antimicrobial resistance of *Campylobacter jejuni* subsp. *jejuni* strains isolated from humans in 1998 to 2001 in Montreal, Canada. *Antimicrob. Agents Chemother.* **47**:2027–2029.

Hakanen, A., P. Kotilainen, P. Huovinen, H. Helenius, and A. Siitonen. 2001. Reduced fluoroquinolone susceptibility in *Salmonella enterica* serotypes in travelers returning from Southeast Asia. *Emerg. Infect. Dis.* **7**:996–1003.

Harbottle, H., S. Thakur, S. Zhao, and D. G. White. 2006. Genetics of antimicrobial resistance. *Anim. Biotechnol.* **17**:111–124.

Helms, M., P. Vastrup, P. Gerner-Smidt, and K. Molbak. 2002. Excess mortality associated with antimicrobial drug-resistant *Salmonella* Typhimurium. *Emerg. Infect. Dis.* **8**:490–495.

Holmstrom, K., S. Graslund, A. Aahlstrom, S. Poungshompoo, B. Bengtsson, and N. Kautsky. 2003. Antibiotic use in shrimp farming and implications for environmental impacts and human health. *Int. J. Food Sci. Technol.* **38**:255–266.

Hummel, R., H. Tschape, and W. Witte. 1986. Spread of plasmid-mediated nourseothricin resistance due to antibiotic use in animal husbandry. *J. Basic Microbiol.* **26**:461–466.

Kassenborg, H. D., K. E. Smith, D. J. Vugia, T. Rabatsky-Ehr, M. R. Bates, M. A. Carter, N. B. Dumas, M. P. Cassidy, N. Marano, R. V. Tauxe, and F. J. Angulo. 2004. Fluoroquinolone-resistant *Campylobacter* infections: eating poultry outside of the home and foreign travel are risk factors. *Clin. Infect. Dis.* **38**(Suppl 3):S279–S284.

Khachatourians, G. G. 1998. Agricultural use of antibiotics and the evolution and transfer of antibiotic-resistant bacteria. *Can. Med. Assoc. J.* **159**:1129–1136.

Kiessling, C. R., M. B. Buen, W. M. Kiessling, E. W. Laster, M. H. Loftis, M. A. J. Jackson, and J. N. Sofos. 2006. Antimicrobial resistance patterns of *Salmonella* isolated from FDA regulated products during 2005. Presented at the FDA Science Forum, 18 to 20 April 2006, Washington, DC.

Kiessling, C. R., M. H. Loftis, W. M. Kiessling, M. B. Buen, E. W. Laster, and J. N. Sofos. 2005. Antibiotic resistance of *Salmonella* isolated from various products, 2003–2004. Presented at the FDA Science Forum, 27 and 28 April 2005, Washington, DC.

Koones, B. 2006. Seafood safety: down on the farm. *Food Safety Mag.* **7**:22–29.

Leegaard, T. M., D. A. Caugnat, L. O. Froholm, E. A. Hoiby, and J. Lassen. 2000. Emerging antibiotic resistance in *Salmonella* Typhimurium in Norway. *Epidemiol. Infect.* **125**:473–480.

Levy, S. B., and B. Marshall. 2004. Antibacterial resistance worldwide: causes, challenges and responses. *Nat. Med.* **10**:S122–S129.

Lin, C. K. 1989. Shrimp culture in Taiwan: what went wrong? *World Aquacult.* **20**:19–20.

Martin, L. J., M. Fyfe, K. Dore, J. A. Buxton, F. Pollari, B. Henry, D. Middleton, R. Ahmed, F. Jamieson, B. Ciebin, S. A. McEwen, and J. B. Wilson. 2004. Increased burden of illness associated with antimicrobial-resistant *Salmonella enterica* serotype Typhimurium infections. *J. Infect. Dis.* **189**:377–384.

McDermott, P. F., S. Zhao, D. D. Wagner, S. Simjee, R. D. Walker, and D. G. White. 2002. The food safety perspective of antibiotic resistance. *Anim. Biotechnol.* **13**:71–84.

Mead, P. S., L. Slutsker, V. Dietz, L. F. McCaig, J. S. Bresee, C. Shapiro, P. M. Griffin, and R. V. Tauxe. 1999. Food-related illness and death in the United States. *Emerg. Infect. Dis.* **5**:607–625.

Molbak, K. 2004. Spread of resistant bacteria and resistance genes from animals to humans—the public health consequences. *J. Vet. Med. B Infect. Dis. Vet. Public Health* **51**:364–369.

Murray, A., J. E. Coia, H. Mather, and D. J. Brown. 2005. Ciprofloxacin resistance in non-typhoidal *Salmonella* serotypes in Scotland, 1993-2003. *J. Antimicrob. Chemother.* **56**:110–114.

Nachamkin, I., H. Ung, and M. Li. 2002. Increasing fluoroquinolone resistance in *Campylobacter jejuni*, Pennsylvania, USA,1982–2001. *Emerg. Infect. Dis.* **8**:1501–1503.

National Research Council. 1998. *The Use of Drugs in Food Animals: Benefits and Risks.* National Academy Press, Washington, DC.

Petersen, A., J. S. Andersen, T. Kaewmak, T. Somsiri, and A. Dalsgaard. 2002. Impact of integrated fish farming on antimicrobial resistance in a pond environment. *Appl. Environ. Microbiol.* **68**:6036–6042.

Phillips, I., M. Casewell, T. Cox, B. De Groot, C. Friis, R. Jones, C. Nightingale, R. Preston, and J. Waddell. 2004. Does the use of antibiotics in food animals pose a risk to human health? A critical review of published data. *J. Antimicrob. Chemother.* **53**:28–52.

Piddock, L. J. 1995. Mechanisms of resistance to fluoroquinolones: state-of-the-art 1992–1994. *Drugs* **49**(Suppl 2):29–35.

Poros-Gluchowska, J., and Z. Markiewicz. 2003. Antimicrobial resistance of *Listeria monocytogenes*. Acta Microbiol. Pol. **52**:113-129.

Prats, G., B. Mirelis, T. Llovet, C. Munoz, E. Miro, and F. Navarro. 2000. Antibiotic resistance trends in enteropathogenic bacteria isolated in 1985–1987 and 1995–1998 in Barcelona. *Antimicrob. Agents Chemother.* **44**:1140–1145.

Prescoot, J. F., J. D. Baggot, and R. D. Walker. 2000. *Antimicrobial Therapy in Veterinary Medicine.* Iowa State University Press, Ames, IA.

Rankin, S. C., H. Aceto, J. Cassidy, J. Holt, S. Young, B. Love, D. Tewari, D. S. Munro, and C. E. Benson. 2002. Molecular characterization of cephalosporin-resistant *Salmonella enterica* serotype Newport isolates from animals in Pennsylvania. *J. Clin. Microbiol.* **40**: 4679–4684.

Reilly, P. J., and D. R. Twiddy. 1992. *Salmonella* and *Vibrio cholerae* in brackishwater cultured tropical prawns. *Int. J. Food Microbiol.* **16**:293–301.

Roberts, M. C., B. Facinelli, E. Giovanetti, and P. E. Varaldo. 1996. Transferable erythromycin resistance in *Listeria* spp. isolated from food. *Appl. Environ. Microbiol.* **62**:269–270.

Schmidt, A. S., M. S. Bruun, I. Dalsgaard, K. Pedersen, and J. L. Larsen. 2000. Occurrence of antimicrobial resistance in fish-pathogenic and environmental bacteria associated with four danish rainbow trout farms. *Appl. Environ. Microbiol.* **66**:4908–4915.

Serrano, P. H. 2005. *Responsible Use of Antibiotics in Aquaculture.* Food & Agriculture Organization, Rome, Italy.

Shen, Y., Y. Liu, Y. Zhang, J. Cripe, W. Conway, J. Meng, G. Hall, and A. A. Bhagwat. 2006. Isolation and characterization of *Listeria monocytogenes* isolates from ready-to-eat foods in Florida. *Appl. Environ. Microbiol.* **72:**5073–5076.

Skov, M., J. Andersen, S. Aabo, S. Ethelberg, F. Aarestrup, and A. Sørensen. 2007. Antimicrobial drug resistance of *Salmonella* isolates from meat and humans, Denmark. *Emerg. Infect. Dis.* **13:**638–641.

Smith, K. E., J. M. Besser, C. W. Hedberg, F. T. Leano, J. B. Bender, J. H. Wicklund, B. P. Johnson, K. A. Moore, and M. T. Osterholm. 1999. Quinolone-resistant *Campylobacter jejuni* infections in Minnesota, 1992–1998. Investigation Team. *N. Engl. J. Med.* **340:** 1525–1532.

Threlfall, E. J., M. Day, E. de Pinna, A. Charlett, and K. L. Goodyear. 2006. Assessment of factors contributing to changes in the incidence of antimicrobial drug resistance in *Salmonella enterica* serotypes Enteritidis and Typhimurium from humans in England and Wales in 2000, 2002 and 2004. *Int. J. Antimicrob. Agents* **28:**389–395.

Threlfall, E. J., J. A. Frost, L. R. Ward, and B. Rowe. 1996. Increasing spectrum of resistance in multiresistant *Salmonella typhimurium*. *Lancet* **347:**1053–1054.

Threlfall, E. J., L. R. Ward, J. A. Frost, and G. A. Willshaw. 2000. The emergence and spread of antibiotic resistance in food-borne bacteria. *Int. J. Food Microbiol.* **62:**1–5.

Tollefson, L., and M. A. Miller. 2000. Antibiotic use in food animals: controlling the human health impact. *J. AOAC Int.* **83:**245–254.

Varma, J. K., K. D. Greene, J. Ovitt, T. J. Barrett, F. Medalla, and F. J. Angulo. 2005a. Hospitalization and antimicrobial resistance in *Salmonella* outbreaks, 1984–2002. *Emerg. Infect. Dis.* **11:**943–946.

Varma, J. K., K. Molbak, T. J. Barrett, J. L. Beebe, T. F. Jones, T. Rabatsky-Ehr, K. E. Smith, D. J. Vugia, H. G. Chang, and F. J. Angulo. 2005b. Antimicrobial-resistant nontyphoidal *Salmonella* is associated with excess bloodstream infections and hospitalizations. *J. Infect. Dis.* **191:**554–561.

Wassenaar, T. M. 2005. Use of antimicrobial agents in veterinary medicine and implications for human health. Crit. Rev. Microbiol. **31:**155–169.

Wegener, H. C. 1999. The consequences for food safety of the use of fluoroquinolones in food animals. *N. Engl. J. Med.* **340:**1581–1582.

White, D. G., A. Datta, P. McDermott, S. Friedman, S. Qaiyumi, S. Ayers, L. English, S. McDermott, D. D. Wagner, and S. Zhao. 2003. Antimicrobial susceptibility and genetic relatedness of *Salmonella* serovars isolated from animal-derived dog treats in the USA. *J. Antimicrob. Chemother.* **52:**860–863.

White, D. G., S. Zhao, R. Singh, and P. F. McDermott. 2004. Antimicrobial resistance among gram-negative foodborne bacterial pathogens associated with foods of animal origin. *Foodborne Pathog. Dis.* **1:**137–152.

Witte, W. 2000a. Ecological impact of antibiotic use in animals on different complex microflora: environment. *Int. J. Antimicrob. Agents* **14:**321–325.

Witte, W. 2000b. Selective pressure by antibiotic use in livestock. *Int. J. Antimicrob. Agents* **16**(Suppl 1):S19–S24.

World Health Organization (WHO). 1997. The medical impact of antimicrobial use in food animals. Report of a WHO meeting. Berlin, Germany, 13–17 October 1997. Berlin, Germany. WHO/EMC/ZOO/97.4. [Online.] http://whqlibdoc.who.int/hq/1997/WHO_EMC_ZOO_97.4.pdf. Accessed 12 November 2007.

World Health Organization (WHO). 2002. Monitoring antimicrobial usage in food animals for the protection of human health: report of a WHO consultation. Oslo, Norway. WHO/CDS/CSR/EPH/2002.11. [Online.] http://whqlibdoc.who.int/hq/2002/WHO_CDS_CSR_EPH_2002.11.pdf. Accessed 12 November 2007.

World Health Organization (WHO). 2006. Antimicrobial use in aquaculture and antimicrobial resistance. Report of a Joint FAO/OIE/WHO Expert Consultation, Seoul, Republic of Korea, 13–16 June 2006. [Online.] http://www.who.int/topics/foodborne_diseases/aquaculture_rep_13_16june2006%20.pdf. Accessed 12 November 2007.

Zhang, Y., E. Yeh, G. Hall, J. Cripe, A. A. Bhagwat, and J. Meng. 2007. Characterization of *Listeria monocytogenes* isolated from retail foods. *Int. J. Food Microbiol.* **113**:47–53.

Zhao, S., S. Qaiyumi, S. Friedman, R. Singh, S. L. Foley, D. G. White, P. F. McDermott, T. Donkar, C. Bolin, S. Munro, E. J. Baron, and R. D. Walker. 2003a. Characterization of *Salmonella enterica* serotype Newport isolated from humans and food animals. *J. Clin. Microbiol.* **41**:5366–5371.

Zhao, S., A. R. Datta, S. Ayers, S. Friedman, R. D. Walker, and D. G. White. 2003b. Antimicrobial-resistant *Salmonella* serovars isolated from imported foods. *Int. J. Food Microbiol.* **84**:87–92.

Zhao, S., P. F. McDermott, S. Friedman, S. Qaiyumi, J. Abbott, C. Kiessling, S. Ayers, R. Singh, S. Hubert, J. Sofos, and D. G. White. 2006. Characterization of antimicrobial-resistant *Salmonella* isolated from imported foods. *J. Food Prot.* **69**:500–507.

Zhao, S., D. G. White, P. F. McDermott, S. Friedman, L. English, S. Ayers, J. Meng, J. J. Maurer, R. Holland, and R. D. Walker. 2001. Identification and expression of cephamycinase bla(CMY) genes in *Escherichia coli* and *Salmonella* isolates from food animals and ground meat. *Antimicrob. Agents Chemother.* **45**:3647–3650.

Imported Foods: Microbiological Issues and Challenges
Edited by Michael P. Doyle and Marilyn C. Erickson
© 2008 ASM Press, Washington, DC

Mycotoxin Contamination of Foods from around the World

7

Garnett E. Wood

INTRODUCTION

Mycotoxins are naturally occurring, toxic secondary metabolites produced by certain fungal species that are present in the soil worldwide. Under favorable conditions of temperature and moisture these fungi are capable of infecting and growing on susceptible agricultural crops in the field, resulting in the production of mycotoxins. The extent of contamination of various food crops by these toxins is greatly influenced by environmental factors such as temperature, humidity, and the extent of rainfall during the growth, harvesting, and postharvesting periods. If the harvested crops are not dried and processed properly, fungi accompanying the crops from the field may continue to grow and produce more toxins, depending on the storage conditions. The worldwide occurrence of mycotoxins in many food products is well documented (Jelinek et al., 1989; Council of Agricultural Science and Technology [CAST], 2003). The extent of mycotoxin contamination that may occur in a given food crop is unpredictable and may vary from year to year, with geographic location and agricultural/agronomic practices in various countries. Hundreds of mycotoxins have been identified in many parts of the world. However, information about many of them is limited with regard to their natural occurrence and stability in foods. It is known that some mycotoxins may exhibit various toxicological manifestations, including teratogenic, mutagenic, and/or carcinogenic effects in susceptible animal species, livestock, and humans. Many mycotoxins are relatively stable compounds in that they are not completely removed or destroyed by many processing techniques, including milling, cooking, and

GARNETT E. WOOD, Office of Food Safety, Center for Food Safety and Applied Nutrition, Food and Drug Administration, College Park, MD 20740.

various thermal procedures. Small amounts of these toxins may be legally permitted in foods in some countries, provided the amounts involved are not considered to be injurious to human health. However, because mycotoxins present a potential public health hazard to humans, exposure to these toxins should be limited.

The U.S. FDA is responsible for enforcing the Federal Food, Drug and Cosmetic Act (the Act) that was passed by Congress. This statute and its amendments serve as the legal basis for the Agency to regulate undesirable substances in foods. Under Section 402(a) (1) of the Act, a food is considered to be adulterated if it contains an added poisonous or deleterious substance that may render it injurious to health. Mycotoxins are considered added poisonous or deleterious substances because their presence in food can be avoided, in part, by the implementation of good agronomic and manufacturing practices. With the exception of most meat and poultry products, which are regulated by the U.S. Department of Agriculture (USDA), all foods that are offered for import into the United States are subject to examination by the FDA and are required to meet the same quality standards as domestic food products. Foods containing poisonous or deleterious substances, such as mycotoxins, that may pose a health hazard, are deemed to be adulterated and subject to enforcement action (e.g., seizure of domestic foods in interstate commerce and detention or refusal of entry for imported foods) under the Act.

The general strategies used by the FDA to minimize mycotoxins in the food supply include (i) establishing guidelines (e.g., action levels, guidance levels) for specific mycotoxins in foods, (ii) monitoring the domestic and import food supply through a formal compliance program to ensure compliance with established guidelines, (iii) providing guidance to the food industry, and (iv) cooperating with state, federal, and international agencies and various trade associations regarding the safety of foods. Action levels are established and used as a guide by the FDA field staff to determine when it may be necessary to take enforcement action against a food producer, processor, or distributor of food products. Guidance levels are guideline levels provided to food producers by the FDA that represent the best estimation of the negligible risk level associated with the presence of a particular mycotoxin in foods based on available exposure and toxicological information. While guidance levels are not enforceable, under the Act, FDA can take appropriate regulatory actions when levels exceeding guideline levels may pose a health hazard. In instances where there are no established guidelines for a particular mycotoxin, a decision to take enforcement action is determined by the Agency on a case-by-case basis.

MONITORING PROGRAM

A significant amount of food consumed in the United States each year is produced, processed, and imported from other countries. Traditional methods of producing, harvesting, and processing raw food crops susceptible to mycotoxin contamination in various countries can result in contamination of some food products that are exported to the United States. Since the history (production, processing procedures, etc.) associated with the foods offered for import into the United States is not known, it is essential that a monitoring program for mycotoxins be in place. The FDA has an ongoing formal compliance program titled "Mycotoxins in Domestic and Imported Foods." This program involves routinely monitoring the food industry. The major objectives of the Compliance Program are to (i) collect and analyze domestic and imported samples of various food products to determine the occurrence and levels of specific mycotoxins of regulatory concern to the Agency, (ii) take appropriate enforcement action against distributors, manufacturers, or importers of foods that contain mycotoxins at levels judged to be of regulatory significance, and (iii) provide a database to support FDA's positions on international issues involving mycotoxins in foods. Variations in levels of mycotoxin contamination resulting from climatic conditions from year to year necessitate monitoring the food industry on a regular basis.

The Compliance Program administered by the FDA's Center for Food Safety and Applied Nutrition (CFSAN) is focused on commodities that historically have been known to be susceptible to mycotoxin contamination or in response to new information on mycotoxin contamination problems developing in various countries. All FDA field offices involved in monitoring activities are provided with a list of commodities susceptible to contamination by specific mycotoxins, a sampling plan, and a quota of samples to be collected each year. Although the FDA compliance program is not statistically designed, inspectors are instructed to collect no more than two samples of any specific commodity for analysis for a given mycotoxin at any location, unless there is a specific need to collect more samples of that commodity for compliance purposes.

The collected samples are analyzed by FDA district laboratories using official, collaboratively studied methods that are found in the current edition of *Official Methods of Analysis of the Association of Official Analytical Chemists International* (AOAC, 2006). Imported food products found not to comply with FDA guideline levels are subject to detention; they must be brought into compliance by the importer based on specific criteria, destroyed, or reexported. Over the years, cooperative monitoring efforts involving the FDA, USDA, and the food industry have resulted in the development of a relatively effective control program for reducing the levels of mycotoxins in domestic and imported foods.

This chapter will focus on the occurrence of the following five mycotoxins in foods imported into the United States during the years 2000 to 2006: aflatoxins, patulin, deoxynivalenol, fumonisins, and ochratoxin A. All of these mycotoxins have been found worldwide in foods and are of regulatory concern to the FDA. Monitoring data obtained by the FDA on the occurrence of these mycotoxins in domestic U.S. foods from earlier years have been published (Pohland and Wood, 1991; Wood and Trucksess, 1998).

MYCOTOXINS IN IMPORTED FOODS

Aflatoxins

The aflatoxins are a group of structurally related toxic compounds produced as secondary metabolites by certain strains of the fungi *Aspergillus flavus*, *Aspergillus parasiticus* and *Aspergillus nomius*. These fungi are commonly found in the soil in tropical and subtropical areas of the world. Under favorable conditions of temperature and humidity, these fungi can infect and grow on susceptible crops in the field and during storage, resulting in the production of aflatoxins. The aflatoxins of major concern are referred to as aflatoxins B_1, B_2, G_1, G_2, and M_1. The first four are usually found together in various foods in varying proportions; aflatoxin B_1 is usually predominant in amount and is the most toxic. Aflatoxin M_1 is a toxic metabolite of aflatoxin B_1. It is produced in the liver of mammals that have ingested a significant amount of aflatoxin B_1, and it is excreted in the milk of dairy cattle, humans, and other mammalian species. The aflatoxins B_1, B_2, G_1, and G_2 have been found occurring naturally in corn and corn products, peanuts and peanut products, various tree nuts, edible seeds, wheat, sorghum, spices, rice, figs, and dried fruit. They have been found sporadically in other commodities at comparatively low or insignificant levels (Wood et al., 2003). The aflatoxins are relatively stable compounds. Varying amounts of aflatoxins may be found in processed food products, depending on the processing procedures employed. They have been shown to be potent hepatocarcinogens in most animal species studied. The International Agency for Research on Cancer (IARC) classified the naturally occurring aflatoxins as being carcinogenic to humans, group 1A; aflatoxin M_1 was classified as 2B, indicating limited evidence for carcinogenicity to humans (IARC, 1993a). These classifications were reaffirmed by IARC in 2002 (IARC, 2002a). A risk assessment on aflatoxins was conducted by the FAO/WHO Joint Expert Committee on Food Additives (JECFA) in 1997 (JECFA, 1998). The Committee, after reviewing results from available scientific and epidemiologic studies involving

aflatoxins, concluded that they should be treated as carcinogenic food contaminants and their intake in humans should be reduced to levels as low as
reasonably achievable.

In view of the possible health hazards associated with aflatoxins, the FDA
established an informal action level of 30 μg/kg for total aflatoxins ($B_1 + B_2 +$
$G_1 + G_2$) in peanut and peanut products in 1965. This level was reduced in
1969 to 20 μg/kg total aflatoxins, and it currently applies to all food commodities (import as well as domestic) that are susceptible to aflatoxin contamination (FDA, 2000). Many countries have attempted to limit exposure to
aflatoxins by the establishment of regulatory limits for foods in domestic and
commercial import channels. As of 2003, 61 countries had maximum toler-
ated levels of 1 to 20 μg/kg for aflatoxin B_1 only in foodstuffs and 76 countries had levels of 0 to 35 μg/kg for total aflatoxins in foodstuffs (FAO,
2004). Many of the countries belonging to the European Community have
harmonized their limits for aflatoxin B_1 and the sum of aflatoxins B_1, B_2, G_1,
and G_2 for various commodities. The United States does not have a separate
limit for aflatoxin B_1 only.

The FDA has a Memorandum of Understanding (MOU) with the USDA
concerning the examination of raw, unprocessed peanuts for aflatoxins. Under
this MOU, the USDA is responsible for the sampling and testing of all imported and domestic raw peanuts for aflatoxins before they are introduced
into commercial channels in the United States. The imported peanuts must
meet the same requirements as domestic peanuts. The importers of peanuts
must offer each lot of raw peanuts to a USDA- or a Peanut Administrative
Committee (PAC)-approved laboratory for testing before introducing that lot
into U.S. commerce. The FDA is responsible for examining imported and domestic roasted peanuts (shelled and in-shell) and all processed peanut products. The FDA also has a Voluntary Control Agreement with the USDA and
importers of Brazil and pistachio raw nuts that has been in existence since
1968 (Stoloff, 1977). Under this agreement, imported in-shell Brazil nuts and
in-shell pistachio nuts are sampled and examined by the USDA, at the importer's expense, for aflatoxins in accordance with procedures prescribed by
the FDA. The FDA is responsible for examining imported and domestic
roasted pistachios and shelled and/or roasted Brazil nuts and their processed
products.

A total of 3,883 samples of imported foods were examined for aflatoxins
by the FDA during the years 2000 to 2006. The breakdown of the data by
type of consumer product is shown in Table 1. In total, 12.5% of the imported food products examined over the 7-year period was contaminated
with aflatoxins; 4.5% contained aflatoxins at levels that exceeded the U.S. action level of 20 μg/kg. The occurrence and levels of contamination were sig-

Table 1 Aflatoxins in foods imported into the United States between 2000 and 2006

| Product | Number analyzed | Aflatoxin levels (μg/kg)[a] | | | | Maximum level |
| | | >1 | | >20 | | |
		No.	%	No.	%	
Bread/bakery	259	21	8.1	9	3.5	175
Breakfast cereals	97	0				
Candy (assorted)	515	136	26.4	48	9.3	740
Corn, canned, dried	108	5	4.6	1	0.9	45
Corn meal	202	0				
Figs/dried fruit	85	10	11.8	5	5.9	140
Snack food	215	24	11.2	5	2.3	102
Spices (assorted)	96	9	9.4	1	1.0	228
Edible seeds						
Lotus	12	2	16.7	1	8.3	82
Melon	156	18	11.5	12	7.7	113
Misc.	118	10	8.5	3	2.5	53
Pumpkin	297	28	9.4	10	3.4	51
Sesame	206	5	2.4	0	0	11
Sunflower	79	2	2.5	0	0	9
Ground nuts						
Peanuts, shelled	180	45	25.0	16	8.9	401
Peanuts, in-shell	84	16	19.0	13	15.5	359
Peanut products	245	78	31.8	14	5.7	668
Tree nuts						
Almonds	73	5	6.8	1	1.4	28
Brazil nuts	70	8	11.4	2	2.8	38
Chestnuts, cashews, macadamia	453	0				
Filberts/hazels	32	3	9.4	1	3.1	71
Mixed	56	4	7.1	0	0	8
Pecans	67	8	11.9	0	0	12
Pine nuts	66	12	18.2	6	9.0	50
Pistachio nuts	86	36	41.9	25	29	211
Walnuts	26	2	7.7	0	0	2

[a] Limit of detection, 1 μg/kg aflatoxin.

nificant in samples of bread/ bakery products, assorted candies, canned corn, edible seeds, figs and dried fruit, peanut and peanut products, tree nuts, snack foods, and spices. The highest level of contamination (740 μg/kg) was found in a sample of candy that contained nuts. A large number of samples of assorted candies examined contained edible seeds, fruit, peanuts, and tree nuts as ingredients; some were chocolate and/or chocolate covered. None of the breakfast cereals, corn meal, chestnuts, and cashew and macadamia nuts examined contained aflatoxins.

Eighty-five samples of figs/dried fruit were examined, and 11.8% were contaminated with aflatoxins; 5.9% at levels greater than the U.S. action level. Aflatoxin contamination in dried figs has been reported in surveys conducted in many countries. It is believed that mycotoxin contamination of figs and other fruit may start with fungal contamination while on the tree, increase during harvesting and drying, and continue to accumulate during storage because of rewetting and improper storage conditions (Drusch and Ragab, 2003). The implementation of good harvesting, drying, and storage practices should reduce the incidence of aflatoxin contamination levels in figs and dried fruit.

Twenty-four of the 215 snack foods examined were contaminated; 2.3% contained aflatoxins at levels greater than the U.S. guideline. The product with the highest level of contamination consisted of a mixture of coconut, rice, and spices as ingredients.

The prevalence of contamination of imported spices was 9.4%, with the highest level of contamination (228 μg/kg) in a sample of mixed spices. Spices are produced mainly in countries with tropical climates. In some countries, it is traditional for spices to be dried on the ground in the open air (Martins et al., 2001). This procedure presents an opportunity for fungi to infect, grow, and produce aflatoxins. Many cultural practices that are implemented in some countries after spices are harvested may help to create ideal conditions for fungal growth and aflatoxin production (Romagnoli et al., 2007).

Of the 868 edible seeds examined, 7.5% were contaminated with aflatoxins. Some of the lotus, melon, pumpkin, and miscellaneous seeds contained aflatoxins at levels greater than the U.S. action level. The highest level of aflatoxin contamination (113 μg/kg) was in a melon seed sample. The incidence and levels of contamination were relatively low in sesame and sunflower seeds. It has been reported that, specifically in some West African countries, high ambient temperatures and humidity provide favorable conditions for the growth of toxigenic fungi in and on edible oilseeds (Bankole et al., 2004). Other factors contributing to aflatoxin contamination include poorly developed processing and storage facilities.

Significant levels of aflatoxin contamination were detected in peanuts and peanut products. Of the 509 products examined, 27.3% contained aflatoxins. The highest incidence and level of contamination were in peanut products, mainly peanut butter; the lowest incidence of contamination was in in-shell peanuts. From surveys conducted in previous years in the United States, it was found that when aflatoxin contamination occurred in peanuts and peanut products, higher incidences and levels were found in peanut butter and shelled peanuts than in in-shell peanuts. A similar trend was observed in surveys of imported products examined between 1982 and 1988 and from 1987 to 1997 (Pohland and Wood, 1991; CAST, 2003). Peanuts are an excellent source of food as well as an excellent substrate for the growth of *A. flavus* and *A. parasiticus* in many countries that experience optimal environmental conditions for growing these molds. It is noteworthy that although *A. flavus* was identified in the early 1960s as the fungus responsible for aflatoxin production in peanuts, now, almost five decades later, aflatoxin contamination in peanuts and peanut products is still of major concern worldwide despite advances in scientific research and technology.

A total of 929 samples of tree nuts were examined, with 78 (8.4%) containing aflatoxins and 35 (3.8%) containing aflatoxins at levels greater than 20 μg/kg. The data in Table 1 reflect the incidence and levels of contamination associated with each type of tree nut examined. Tree nuts containing aflatoxins at levels greater than the FDA action level included almond, Brazil, filbert/hazel, pine, and pistachio with the highest incidence and level of contamination associated with pistachio, nuts. None of the 453 samples of chestnut, cashew, and macadamia nuts contained aflatoxins.

Tree nuts, in general, are exposed to high levels of airborne fungal spores while growing on the tree and also during harvest and/or processing (Bayman et al., 2002). These fungi can colonize and remain with the nuts during processing; therefore, temporary exposure to conditions that favor fungal growth can result in internal infections in the nuts. The primary infection of pistachio nuts with aflatoxin-producing fungi is believed to take place in orchards as a result of insect damage or physical abuse that can occur when nuts are shaken from the trees during harvest or transportation or storage under adverse conditions.

The occurrence of aflatoxins in Brazil nuts has been widely reported in the literature; however, the critical points for infection with aflatoxin-producing fungi resulting in toxin formation have not been elucidated (Marklinder et al., 2005). Basic issues that should be considered by harvesters and processors for the safe handling and storage of Brazil nuts in order to limit aflatoxin production have been published (Arrus et al., 2005).

All of the products examined for aflatoxins during the 3-year period of 2004 to 2006 were imported from 70 countries, including 18 in the Americas, 9 in Africa, 25 in Asia, and 17 in Europe and Australia. Aflatoxins at levels greater than 20 μg/kg were found in some products from Canada, China, El Salvador, India, Iran, Israel, Jordan, Lebanon, Mexico, Pakistan, Philippines, Thailand, Turkey, and Vietnam. These data support previous observations that aflatoxin contamination in foods is a worldwide problem (Jelinek et al., 1989; JECFA, 1998). It was recently reported that on a global scale, approximately 4.5 billion persons living in developing countries are chronically exposed to largely un-controlled amounts of aflatoxins (Williams et al., 2004).

The data in Table 1 do not include imported raw peanuts examined by the USDA as a result of a MOU with FDA. Under this MOU, a total of 273 lots of raw peanuts from Argentina, China, and Nicaragua were examined for aflatoxins in USDA laboratories during 2006; none of the lots contained aflatoxins at levels greater than the U.S. action level of 20 μg/kg. As a result of a Voluntary Control Agreement involving the importers of raw nuts, USDA, and FDA, 96 lots of raw Brazil nuts were examined for aflatoxins by the USDA during 2006; only one lot was contaminated with aflatoxins at a level greater than 20 μg/kg (USDA, privileged communication).

The Codex Alimentarius is the joint food standards program of the Food and Agriculture Organization (FAO) and the World Health Organization (WHO) of the United Nations. This body is composed of approximately 170 countries and was established to protect the health of consumers and to facilitate trade through the establishment of international food standards. The Codex Alimentarius Commission (CAC) adopted a maximum level of 15 μg/kg for total aflatoxins in peanuts for further processing in 1999 (CAC, 1999). More recently, a Code of Practice for the Prevention and Reduction of Aflatoxin Contamination in Peanuts was adopted by the 27th Session of the CAC in 2004 (CAC, 2004). This Code of Practice document contains uniform guidance for all countries to consider in attempting to control afla-toxin contamination in peanuts.

A Code of Practice for the Reduction and Prevention of Aflatoxin Contamination in Tree Nuts was adopted by the CAC in 2005 (CAC, 2006). This Code applies to all varieties of tree nuts of commercial and international concern. The code contained recommended practices based on Good Agricultural Practices (GAPs), Good Manufacturing Practices (GMPs), and Good Storage Practices (GSPs) that should be sanctioned by national au-thorities in all countries involved in the tree nut industry.

Patulin

Patulin is a toxic metabolite produced by fungi of several genera, including *Penicillium, Aspergillus,* and *Byssochylamys.* These fungi can grow on a variety of foods, including fruits. Under natural conditions, patulin has been isolated al-most exclusively from apples and apple products contaminated with *Penicillium expansum.* Fungal spores can infect and grow on diseased or insect-damaged apples on the tree or when fallen fruit is allowed to remain on the ground prior to storage or processing. Patulin is not completely destroyed by thermal pro-cessing and can occur at high levels in apple juice, including pasteurized juice, if rotten, moldy, or damaged apples are used to make the juice. Since damage

Table 2 Patulin in apple juice imported into the United States between 2000 and 2006

Number analyzed	Patulin levels (µg/kg)[a]				Maximum level
	>10		>50		
	No.	%	No.	%	
933	473	50.7	17	1.82	696

[a]Limit of detection, 10 µg/kg.

to the fruit is not easily observed, presence of patulin in juice made from seemingly wholesome apples cannot be totally excluded.

Studies revealed that patulin has immunosuppressive effects in experimental animals. The IARC reviewed available toxicological information available on patulin and concluded that no evaluation could be made of the carcinogenicity of patulin in humans and that there was inadequate evidence in experimental animals (IARC, 1986). Animal feeding studies have demonstrated, however, that high levels of patulin in apple juice could pose a potential health risk to humans, particularly in children (Selmanoglu, 2006).

In 2001, the United States established an action level of 50 µg/kg for patulin in apple juice, apple juice concentrates, and apple juice products based on the level found in single-strength apple juice or in the single-strength apple juice component of a juice product (FDA, 2001a). At least 48 countries now regulate patulin in apple juice at levels of 30 to 50 µg/kg, with lower levels in infant food (FAO, 2004).

Data on patulin levels in samples of apple juice imported into the United States between 2000 and 2006 are shown in Table 2. Nine-hundred and thirty-three samples of apple juice were examined, of which 50.7 % were contaminated with patulin and 1.8 % contained patulin at levels greater than the U.S. action level of 50 µg/kg. The highest level of patulin present was 696 µg/kg. The apple juice samples examined during the 3-year period of 2004 to 2006 were imported from 34 countries, including 11 in the Americas, 1 in Africa, 8 in Asia, and 13 in Europe and New Zealand.

It is possible to control the level of patulin in apple juice to some extent. It has been reported that the occurrence of patulin in apple juice and apple juice products can be greatly reduced if apple processors follow Good Manufacturing Practices of trimming damaged apples and removing spoiled or rotten apples from the processing system used for the production of apple juice (Lovett et al., 1975; Sydenham et al., 1997).

The CAC also adopted a maximum level of 50 µg/kg patulin in apple juice and apple juice used as an ingredient in other beverages, as well as a Code of Practice for the Prevention and Reduction of Patulin Contamination in Apple Juice and Apple Juice Ingredients in Other Beverages in 2003 (CAC, 2003a).

The recommendations for reducing patulin contamination in apple juice in that document were divided into two parts: (i) recommended practices based on Good Agricultural Practices (GAPs), and (ii) recommended practices based on Good Manufacturing Practices (GMPs). Implementation of those recommendations by countries that are members of the CAC should greatly reduce patulin levels in apple juice in international trade.

Deoxynivalenol

Deoxynivalenol (DON), commonly known as vomitoxin, belongs to a class of sesquiterpenoid mycotoxins that are referred to as trichothecenes. The trichothecenes are produced by several fungi of the genus *Fusarium*, especially *F. graminearum* and *F. culmorum*, which are pathogens of wheat, rye, corn, and other cereal grains. These fungi are found worldwide in the soil, and the distribution of the two species appears to be related to temperature, with *F. graminearum* occurring predominately in crops grown in the warmer climates of the world and *F. culmorum* occurring predominately in crops, grown in the cooler climatic regions (Miller et al., 1991). These fungi are also responsible for the *Fusarium* head blight disease (FHB) in cereal crops, resulting in the production of DON. Studies have shown that the severity of FHB infections depends mainly on climatic effects (temperature, rainfall, humidity) (Champeil et al., 2004). It is not possible to completely avoid the presence of DON in cereal grains, specifically wheat. DON is sometimes found in wheat grown under normal weather conditions; however, the fungi survives best when there is cool, wet weather at the silking or anthesis stage of grain development. The extent to which DON is removed from contaminated grains during milling operations can vary, depending on the procedures used. DON is water soluble, relatively heat stable, very stable during storage of grains, and stable to most processing and cooking procedures (Trigo-Stockli, 2002).

In limited surveys conducted in the United States during 1982, it was found that wheat crops in some midwestern states were heavily infected with FHB disease and contained high levels of DON. Limited toxicological studies in experimental animals at that time revealed that high levels of DON caused adverse immunologic and embryo toxic effects. In view of the potential public health hazard associated with these observations, the FDA issued a guidance (advisory) level for DON in wheat and wheat products in 1982. In 1993, the FDA issued a revised guidance level for DON in finished wheat products intended for human consumption. This level was derived as a result of considerations given to available toxicological data, including reports of outbreaks of DON-associated acute gastrointestinal illnesses in humans in China in 1984/1985 and in India in 1987 and other public health concerns. The current U.S. guidance level for DON in finished wheat products

Table 3 Deoxynivalenol in wheat products imported into the United States between 2000 and 2006

Product	Number analyzed	Deoxynivalenol levels (mg/kg)[a]				Maximum level
		>0.02		>1		
		No.	%	No.	%	
Bakery products	20	2	10.0	0	0	0.05–0.5
Breakfast cereal	8	4	50.0	2	25.0	0.72–3.4
Flours, assorted grains	24	9	37.5	1	4.2	0.08–1.2
Wheat bran	14	3	21.4	0	0	0.04–0.5
Wheat flour	248	76	30.6	0	0	0.28–0.6
Wheat meal/other milled products	84	15	17.8	1	1.2	0.16–1.1

[a]Limit of detection, 0.02 mg/kg.

(e.g., flour, bran, and germ) for human food is 1 mg/kg (FDA, 1993). Thirty-seven countries have established regulatory levels ranging from 0 to 2 mg/kg for DON in wheat flour and other cereals as of 2003 (FAO, 2004)

The IARC reviewed in 1993 the toxicity of DON and some other toxins derived from *Fusarium graminearum* and *Fusarium culmorum* (IARC, 1993b). The IARC concluded that there was inadequate evidence in humans for the carcinogenicity of toxins derived from *Fusarium graminearum*, that there was inadequate evidence in experimental animals for the carcinogenicity of DON, and that the toxins derived from *F. graminearum* and *F. culmorum* were not classifiable as to their carcinogenicity to humans (Group 3). A review of the toxicity of DON and its potential effects on humans has been published recently (Pestka and Smolinski, 2005). The primary human safety concern for DON is its potential for inducing acute gastroenteritis with vomiting.

Data on DON levels in foods imported into the United States during 2000 to 2006 are shown in Table 3. Twenty-seven percent of the 398 products examined contained DON with 1% containing levels in excess of 1 mg/kg. Some DON contamination was present in each food type analyzed. The highest level of DON detected was 3.40 mg/kg in a sample of breakfast cereal. None of the bakery products, wheat bran, and wheat flour samples contained DON at levels in excess of the U.S. guidance level of 1 mg/kg. The products examined for DON during the 3-year period of 2004 to 2006 were imported from 33 countries including 4 in the Americas, 2 in Africa, 14 in Asia, and 12 in Europe and Australia.

Fumonisins

The fumonisins are toxic metabolites produced mainly by the fungi *Fusarium verticillioides* (formerly *F. moniliforme*), *F. proliferatum*, and other *Fusarium* species that can grow on certain crops in the field and during storage. *F. verticillioides* is a soil-borne as well as a seed-borne pathogen of corn. The major naturally occurring fumonisins are referred to as fumonisin B_1 (FB_1), fumonisin B_2 (FB_2), and fumonisin B_3 (FB_3). The most prevelant of these mycotoxins is FB_1, which is also believed to be the most toxic (Thiel et al., 1992). The worldwide occurrence of fumonisins in corn and corn-based food products has been well documented (Shephard et al., 1996; Marasas, 1996).

The extent of fumonisin contamination of raw corn varies with geographic location and agronomic practices in different countries. The levels of fumonisins in corn are influenced by environmental factors such as temperature, humidity, and rainfall during the growth and harvesting periods; high levels have been associated with hot, dry weather followed by periods of high humidity (Shelby et al., 1994). Fumonisin levels in harvested corn are also influenced by improper drying and storage conditions. Fumonisins are relatively heat stable; therefore, ordinary cooking and heat processes do not substantially reduce their levels in foods.

Studies have shown that fumonisins have been associated with a variety of adverse health effects in livestock and experimental animals. The IARC classified the toxins derived from *F. verticillioides* as possible human carcinogens (Group 2B) in 1993 (IARC, 1993c). In 2002, after a review of additional data, the IARC concluded that fumonisin B_1 is possibly carcinogenic to humans (Group 2B) (IARC, 2002b). Epidemiologic studies conducted in several countries have revealed that fumonisins are toxic to humans, but their relationship to human esophageal or liver cancer has not been definitely demonstrated (Wood et al., 2003).

To protect the public health of consumers, the FDA issued a guidance document to the food industry in 2001 that represented the Agency's current thinking on the control of fumonisins in human foods (FDA, 2001b). That document recommends fumonisin levels that FDA considers adequate to protect human health and that are achievable in human foods with the use of good agricultural and good manufacturing practices. The following maximum guidance levels have been recommended by the FDA for total fumonisins (FB1 + FB2 + FB3) in human foods: 2 mg/kg in degermed dry milled corn products, 3 mg/kg for clean corn intended for popcorn, and 4 mg/kg in whole or partially degermed dry milled products, dry milled corn bran, and cleaned corn intended for masa production. As of 2003, six countries had established limits for fumonisins in corn ranging from 1 to 3 mg/kg (FAO, 2004).

Data on imported samples examined for fumonisins from 2000 to 2006 are shown in Table 4. Three hundred and forty-one samples of bakery products, canned/dried corn, corn cereals, corn meal, and corn snacks were examined, of which 86 (25.2%) were contaminated with fumonisins. The

Table 4 Fumonisins in corn products imported into the United States between 2000 and 2006

Product	Number analyzed	Fumonisins levels (mg/kg)[a]				Range
		>0.025		>4		
		No.	%	No.	%	
Bakery products	29	7	24.1	1	3.4	0.26–4.06
Corn, canned, dried	55	18	32.7	3	5.4	0.1–5.58
Corn cereals	73	10	13.7	0	0	0.1–2.84
Corn meal	138	39	28.2	0	0	0.04–3.2
Corn snacks	46	12	26.1	0	0	0.1–0.6

[a]Limit of detection, 0.025 mg/kg.

prevalence of contamination was relatively high in each type of product examined; however, only 1.2% of the total products examined exceeded the FDA guidance level of 4 mg/kg for fumonisins in human food. The levels of contamination ranged from 0.04 mg/kg in cornmeal to 5.58 mg/kg in a sample of canned corn. The lower incidences of contamination detected in corn cereals, bakery products, and corn snacks might have resulted from the use of more refined procedures in their manufacturing processes. The products examined during the 3-year period of 2004 to 2006 were imported from 29 countries, including 14 in the Americas, 3 in Africa, 7 in Asia, and 5 in Europe.

The levels of fumonisins in processed corn products for human consumption have been reported to vary depending on the milling (such as wet or dry milling) and manufacturing processes to which raw corn is subjected (Bennett and Richard, 1996). Specifically, milling of fumonisin-contaminated corn reduces the level of toxins in products such as cornmeal and flour and more severe processing procedures involving intense thermal and alkali treatments can further reduce fumonisin levels in other corn products (Park et al., 2004).

The prevalence and levels of fumonisins in corn crops around the world vary considerably depending on many factors, including environmental conditions, extent of insect damage, cultivar of corn planted, and agronomic practices employed. Information currently available on the occurrence of fumonisins in corn suggests that immediate research should be focused on the development of control measures embodied in a GAPs program. Implementation of these practices, along with advances in postharvest techniques involving proper drying and storage conditions followed by GMPs, could substantially reduce the levels of fumonisins in the food supply. In addition, more research is needed regarding the fate of fumonisins during various industrial processing operations.

Ochratoxin A

Ochratoxin A (OTA) is a naturally occurring toxic fungal metabolite produced by certain species of the genera *Aspergillus* and *Penicillium*. These fungi are associated with many food crops that are grown in various parts of the world. Susceptible crops may become infected with these fungi while in the field or during improper storage conditions. OTA is mainly a contaminant of cereals (corn, barley, wheat, and oats) but has also been found in beans (coffee, soya, and cocoa), dried vine fruit, figs, nuts, and spices. The prevalence and levels of OTA in various crops worldwide may vary from year to year between countries and within countries, depending on climatic conditions and agronomic practices. *Aspergillus ochraceus* is reported to be the major producer of OTA in countries with tropical and subtropical climates; *Penicillium verrucosum* is the main producer of OTA in stored cereals in countries with temperate climates (Walker, 2002). *Aspergillus carbonarius* is the species responsible for OTA contamination of grapes, wine, and vine fruit (Magan and Alfred, 2005). OTA is a relatively stable molecule and is not completely destroyed by milling and various processing procedures (Osborne et al., 1996).

OTA has been described as nephrotoxic, carcinogenic, teratogenic, immunotoxic, and hepatotoxic in laboratory and domestic animals, and as the probable causal agent of the development of nephropathic diseases in humans in certain Balkan countries (O'Brien and Dietrich, 2005). OTA was classified by IARC as possibly carcinogenic for humans (group 2B) (IARC, 1993d).

No regulatory guideline has been established for OTA by FDA because it has been found (via surveys) that the extent of its natural occurrence in domestic U.S. foods is very low and therefore the possibility of human exposure is limited. Under the Food, Drug, and Cosmetic Act, however, any food containing OTA at a level that may be considered injurious to human health is considered adulterated and therefore may be subjected to enforcement action on a case-by-case basis by the Agency. Although the FDA has not established a guidance level for OTA, any level above 20 ppb is of regulatory concern to the Agency. As of 2003, 37 countries had established regulatory limits for OTA ranging from 3 to 50 μg/kg in cereal and cereal products (FAO, 2004).

Data on OTA levels in imported foods from 2000 to 2006 are shown in Table 5. Of the 943 samples examined, 4.1% were contaminated with OTA, with the highest level being 40.3 μg/kg in a sample of canned beans. None of the bakery products, corn meal and milled products, rice flour, and wheat flour contained OTA. A high prevalence of contamination occurred in oats, raisins, and coffee beans. The products examined for OTA during the 3-year period of 2004 to 2006 were imported from 44 countries, including 17 in the Americas, 4 in Africa, 15 in Asia, and 7 in Europe and Australia.

Table 5 Ochratoxin A in foods imported into the United States between 2000 and 2006

Product	Number analyzed	Ochratoxin A levels (μg/kg)[a] >2 No.	Ochratoxin A levels (μg/kg)[a] >2 %	Maximum level
Bakery products	48	0		
Barley/barley malt	68	3	4.4	7.2
Breakfast cereals	185	4	2.2	3.9
Beans (canned)	193	5	2.6	40.3
Coffee beans	162	17	10.5	7.5
Corn meal/milled products	27	0		
Flour, assorted grains, milled products	104	4	3.8	11.3
Flour, rice	19	0		
Flour, wheat	84	0		
Oats	26	3	11.5	7.7
Raisins	27	3	11.1	20.4

[a]Limit of detection, 2 μg/kg.

High prevalences of OTA contamination in raisins and other dried fruits have been observed in many countries (Lombaert et al., 2004; Zinedine et al., 2007). The extent of OTA contamination in coffee beans can be attributed to climatic conditions, abnormally long storage, and traditional processing procedures implemented in different countries (Romani et al., 2000). It has also been reported that infection of coffee beans occurs mostly after harvest, and the fungal sources are likely to be the soil, equipment, and/or drying procedures (Taniwaki et al., 2003). Coffee beans are often processed outdoors or in semienclosed facilities; therefore, it is difficult to control moisture during processing and storage (Bayman and Baker, 2006).

INTERNATIONAL ACTIVITIES

The FDA organized and cosponsored the International Workshop on Mycotoxins at College Park, Maryland, 22 to 26 July 2002. This workshop was an international outreach effort by the FDA to assist economically challenged nations worldwide in minimizing the problem of mycotoxin contamination of foodstuffs in their respective countries. The long-term objectives of the workshop were to reduce human and animal exposure to mycotoxins through increased (i) awareness of health risks associated with mycotoxin contamination, (ii) accessibility of training and detection methods, (iii) knowledge of conditions leading to mycotoxin formation, (iv) regulation and monitoring programs, and (v) compliance with international trade standards. More than

100 participants attended this workshop, representing 45 countries: 16 from Africa, 10 from the Americas, 10 from Asia, and 9 from Europe. The participants returned to their respective countries with the goal of assessing the necessary elements for establishing and implementing local training workshops in their countries.

Scientists in the FDA's Center for Food Safety and Applied Nutrition have been actively participating in working groups of the CAC on issues involving the worldwide contamination of aflatoxins, patulin, DON, fumonisins, and OTA in various cereal crops. Specifically, working through the Codex Committee on Contaminants in Foods (formerly Codex Committee on Food Additives and Contaminants), they have been heavily involved in drafting codes of practices documents and providing discussion papers designed to focus attention on the prevention and reduction of contamination in the food supply throughout the world. Those scientists, with the assistance of scientists from six countries, were the lead drafters of a document titled "A Code of Practice for the Prevention and Reduction of Mycotoxin Contamination in Cereals, Including Annexes on Ochratoxin A, Zearalenone, Fumonisins and Trichothecenes" (CAC, 2003b). That document was adopted by the CAC in 2003 and contains general principles and practices for the reduction of various mycotoxins in cereals that should be initiated by various countries. The implementation of suggested practices pointed out in that document, along with advances in postharvest techniques, proper drying and storage conditions followed by GMPs, should substantially reduce in various countries the levels of the major mycotoxins of concern in the human food supply.

CONCLUSIONS

The monitoring data obtained on food products offered for import into the United States between 2000 and 2006 revealed that a significant number of products were frequently contaminated with mycotoxins of regulatory concern to the FDA. In view of our current knowledge of how food can become contaminated with mycotoxins in industrialized as well as developing countries, one cannot expect all imported foods to be free of mycotoxins. The extent of mycotoxin contamination of foods is greatly influenced, in part, by uncontrollable environmental conditions that prevail during the growth, harvesting, and processing of crops that are used for the preparation of human food.

Some foreign industries, in particular, in developing countries, continue to resort to traditional agronomic and food processing procedures that have been followed for many years. In some countries, the lack of adequate resources prevents them from being able to keep abreast of modern technological advances in the harvesting, processing, and testing of food products designated for export to other countries.

The action levels and guidance levels established by the FDA, coupled with a formal compliance program for foods, help to reduce the risk of mycotoxin exposure from imported foods by preventing the entry of such food into domestic food channels. Implementation of current regulatory programs for mycotoxins in the United States will continue until advances in research and/or advanced processing procedures result in further reduction in the mycotoxin content of foods in domestic and international channels. The enforcement of regulatory limits on mycotoxins imposed by importing countries should be an incentive for all countries involved in international trade to implement effective control measures in their country. There is an awareness that some of the strategies effective for reducing mycotoxin contamination in food in developed countries cannot be implemented in developing countries for a variety of reasons, including climatic conditions as well as the many economic challenges present. Suggested control measures that could be implemented in a developing country like West Africa have been published (Bankole and Adelbanjo, 2003).

REFERENCES

Association of Official Analytical Chemists (AOAC). 2006. *Official Methods of Analysis of AOAC International*, 18th ed. Association of Official Analytical Chemists. Gaithersburg, MD. [Online.] http://eoma.aoac.org. Accessed 12 November 2007.

Arrus, F., G. Blank, D. Abramson, R. Clear, and R. A. Holley. 2005. Aflatoxin production by *Aspergillus flavus* in Brazil nuts. *J. Stored Products Res.* 41:513–527.

Bankole, S. A., and A. Adelbanjo. 2003. Mycotoxins in food in West Africa: current situation and possibilities of controlling it. *African J. Biotech.* 2(9):254–263.

Bankole, S. A., B. M. Ogunsanwo, and O. O. Mabekojoe. 2004. Natural occurrence of moulds and aflatoxin B_1 in melon seeds from markets in Nigeria. *Food Chem. Toxicol.* 42:1309–1314.

Bayman, P., and J. L. Baker. 2006. Ochratoxins: a global perspective. *Mycopathologia* **162:** 215–223.

Bayman, P., J. L. Baker, and N.E. Mahoney. 2002. *Aspergillus* on tree nuts: incidence and associations. *Mycopathologia* 155:161–169.

Bennett, G. A., and J. L.Richard. 1996. Influence of processing on *Fusarium* mycotoxins in contaminated grains. *Food Tech.* 50(5):235–238.

Champeil, A., J. F. Fourbet, T. Dore, and L. Rossignol. 2004. Influence of cropping system on *Fusarium* head blight and mycotoxin levels in winter wheat. *Crop Prot.* 23:531–537.

Codex Alimentarius Commission (CAC). 1999. Maximum level of 15 µg/kg for total aflatoxins in peanuts intended for further processing. Codex Alimentarius Commission Twenty-third Session, Rome, Italy, 28 June to 3 July 1999. ALINORM 99/37 Par. 102. [Online.] http://www.fao.org/docrep/meeting/005/X2630E/X2630E01.htm. Accessed 12 November 2007.

Codex Alimentarius Commission (CAC). 2003a. Code of practice for the prevention and reduction of patulin in apple juice and apple juice ingredients in other beverages. CAC/RCP 50-2003. [Online.] www.codexalimentarius.net/download/standards/405/CXC_050e.pdf. Accessed 12 November 2007.

Codex Alimentarius Commission (CAC). 2003b. Code of practice for the prevention and reduction of mycotoxin contamination in cereals, including annexes on ochratoxin A, zearalenone, fumonisins and trichothecenes. CAC/RCP 51-2003. [Online.] www.codexalimentarius.net/download/standards/406/CXC_051e.pdf. Accessed 13 November 2007.

Codex Alimentarius Commission (CAC). 2004. Code of practice for the prevention and reduction of aflatoxin contamination in peanuts. CAC/RCP 55-2004. [Online.] http://www.codexalimentarius.net/download/standards/10084/CXC_055_2004e.pdf. Accessed 13 November 2007.

Codex Alimentarius Commission (CAC). 2006. Code of practice for the prevention and reduction of aflatoxin contamination in tree nuts. CAC/RCP 59-2005, Rev.1-2006. [Online.] http://www.codexalimentarius.net/download/standards/10221/CXP_059e.pdf. Accessed 13 November 2007.

Council of Agricultural Science and Technology (CAST). 2003. Occurrence of mycotoxins in food and feed, p.36-47. *In*: J. L. Richard and G.A. Payne (ed.), *Mycotoxins: Risks in Plant, Animal and Human Systems.* CAST Task Force Report no. 139. Council of Agricultural Science and Technology, Ames, Iowa.

Drusch, S., and W. Ragab. 2003. Mycotoxins in fruits, fruit juices, and dried fruit. *J. Food Prot.* **66:**1514–1527.

Food and Agriculture Organization (FAO). 2004. Worldwide regulations for mycotoxins in food and feed in 2003, p. 1–165. FAO Food and Nutrition Paper 81, Rome, Italy. ISBN 92-5-105162-3.

Food and Drug Administration (FDA). 1993. Guidance for Industry and FDA. Letter to state agricultural directors, state feed control officials, and food, feed, and grain trade organizations. [Online.] http://www.cfsan.fda.gov/~dms/graingui.html. Accessed 13 November 2007.

Food and Drug Administration (FDA). 2000. Action levels for added poisonous or deleterious substances in human food and animal feed. [Online.] http://vm.cfsan.fda.gov/~lrd/fdaact.html. Accessed 13 November 2007.

Food and Drug Administration (FDA). 2001a. Patulin in apple juice, apple juice concentrates and apple juice products. [Online.] http://www.cfsan.fda.gov/~dms/patubck2.html. Accessed 13 November 2007.

Food and Drug Administration (FDA). 2001b. Guidance for industry. Fumonisin levels in human foods and animal feeds. [Online.] http://www.cfsan.fda.gov/~dms/fumongu2.html. Accessed 13 November 2007.

International Agency for Research on Cancer (IARC). 1986. Some naturally occurring toxins and synthetic food components, furocoumarins and ultraviolet radiation: patulin, p. 83–89. IARC *Monographs on the Evaluation of Carcinogenic Risks to Humans,* vol. 40. International Agency for Research on Cancer, Lyon, France.

International Agency for Research on Cancer (IARC). 1993a. Some naturally occurring substances: food items and constituents, heterocyclic aromatic amines and mycotoxins: aflatoxins,

p. 245–395. *IARC Monographs on the Evaluation of Carcinogenic Risks to Humans,* vol. 56. International Agency for Research on Cancer, Lyon, France.

International Agency for Research on Cancer (IARC). 1993b. Some naturally occurring substances: food items and constituents, heterocyclic aromatic amines and mycotoxins. toxins derived from *Fusarium graminearum, F. culmorum,* and *F. crookwellense* : zearalenone, deoxynivalenol, nivalenol and fusarenone X, p. 397–444. *IARC Monographs on the Evaluation of Carcinogenic Risks of Chemicals to Humans,* vol. 56. International Agency for Research on Cancer, Lyon, France.

International Agency for Research on Cancer (IARC). 1993c. Some naturally occurring substances: food items and constituents, heterocyclic aromatic amines and mycotoxins: fumonisins, p. 445–466. *IARC Monographs on the Evaluation of Carcinogenic Risks of Chemicals to Humans,* vol. 56. International Agency for Research on Cancer, Lyon, France.

International Agency for Research on Cancer (IARC). 1993d. Some naturally occurring substances: food items and constituents, heterocyclic aromatic amines and mycotoxins: ochratoxin A, p. 489–521. *IARC Monographs on the Evaluation of Carcinogenic Risks of Chemicals to Humans,* vol.56. International Agency for Research on Cancer, Lyon, France.

International Agency for Research on Cancer (IARC). 2002a. Aflatoxins. Some traditional herbal medicines, some mycotoxins, naphthalene and styrene, p. 171–300. *IARC Monographs on the Evaluation of Carcinogenic Risks to Humans,* vol. 82. International Agency for Research on Cancer, Lyon, France.

International Agency for Research on Cancer (IARC). 2002b. Fumonisin B1. Some traditional herbal medicines, some mycotoxins, naphthalene and styrene, p. 301–366. *IARC Monographs on the Evaluation of Carcinogenic Risks to Humans,* vol. 82. International Agency for Research on Cancer, Lyon, France.

Joint FAO/WHO Expert Committee on Food Additives (JECFA). 1998. Safety evaluation of certain food additives and contaminants: aflatoxins, p. 359–469. 49th Meeting of the Joint FAO/WHO Expert Committee on Food Additives. WHO Food Additive Series No. 40. World Health Organization, Geneva, Switzerland.

Jelinek, C.F., A. E. Pohland, and G.E.Wood. 1989. Worldwide occurrence of mycotoxins in foods and feeds—an update. *J. Assoc. Off. Anal. Chem.* **72:**223–230,

Lombaert, G.A., Pellaers, G. Neumann, D. Kitchen, V. Huzel, R. Trelka, S. Kotello, and P. M. Scott. 2004. Ochratoxin A in dried vine fruits on the Canadian retail market. *Food Addit. Contam.* **21:**578–585.

Lovett, J., R. G. Thompson, and B. Boutin. 1975. Trimming as a means of removing patulin from fungus-rotted apples. *J. Assoc. Off. Anal. Chem.* **58:**909–911.

Magan, N., and D. Aldred. 2005. Conditions of formation of ochratoxin A in drying, transport and in different commodities. *Food Addit. Contam.* **22**(Suppl. 1):10–16.

Marasas, W. F. O. 1996. Fumonisins: history, world-wide occurrence and impact, p. 1–18. *In* L. S. Jackson, J. W. DeVries, and L. B. Bullerman (ed.), *Fumonisins in Food.* Plenum Press, New York, N. Y.

Marklinder, I., M. Lindblad, A. Gidlund, and M. Olsen. 2005. Consumer's ability to discriminate aflatoxin-contaminated Brazil nuts. *Food Addit. Contam.* **22:**56–64.

Martins, M. L., H. M. Martins, and F. Bernardo. 2001. Aflatoxins in spices marketed in Portugal. *Food Addit. Contam.* **18:**315–319.

Miller, J. D., R. Greenhalgh, Y.-Z. Wang, and M. Lu. 1991. Trichothecene chemotypes of three *Fusarium* species. *Mycologia* **83**:121–130.

O'Brien, E., and D. R. Dietrich. 2005. Ochratoxin A: the continuing enigma. *Crit. Rev. Toxicol.* **35**:33–60.

Osborne, B. G., F. Ibe, G. L. Brown, F. Petagine, K. A. Scudamore, J. N. Banks, M. T. Helmanski, and C. T. Leonard. 1996. The effects of milling and processing on wheat contaminated with ochratoxin A. *Food Addit. Contam.* **13**:141–153.

Park, J. W., P. M. Scott, B. P.-Y Lau, and D. A. Lewis. 2004. Analysis of heat-processed corn foods for fumonisins and bound fumonisins. *Food Addit. Contam.* **21**:1168–1178.

Pestka, J. J., and A. T. Smolinski. 2005. Deoxynivalenol: toxicology and potential effects on humans. *J. Toxicol. Environ. Health* **8**(Part B):39–69.

Pohland, A. E., and G. E. Wood. 1991. Natural occurrence of mycotoxins, p. 32–52. *In* G. A. Bray and D. H. Ryan (ed.) *Mycotoxins, Cancer, and Health.* Louisiana State University Press, Baton Rouge, La.

Romagnoli, B., V. Menna, N. Gruppioni, and C. Bergamini. 2007. Aflatoxins in spices, aromatic herbs, herb-teas and medicinal plants marketed in Italy. *Food Control* **18**:697–701.

Romani, S., G. Sacchetti, C. C. Lopez, G. G. Pinnavaia, and M. D. Rosa. 2000. Screening on the occurrence of ochratoxin A in green coffee beans of different origins and types. *J. Agric. Food Chem.* **48**:3616–3619.

Selmanoglu, G. 2006. Evaluation of the reproductive toxicity of patulin in growing male rats. *Food Chem. Toxicol.* **44**:2019–2024.

Shelby, R. A., D. G. White, and E. M. Bauske. 1994. Differential fumonisin production in maize hybrids. *Plant Dis.* **78**:582–584.

Shephard, G.S., P. G. Thiel, S. Stockenstrom, and E. W. Sydenham. 1996. Worldwide survey of fumonisin contamination of corn and corn-based products. *J. AOAC Int.* **79**: 671–687.

Stoloff, L. 1977. Aflatoxins-an overview, p. 7–28. *In* J. V. Rodricks, C.W. Hesseltine, and M. A. Mehlman (ed.), *Mycotoxins in Human and Animal Health.* Pathotox, Park Forest South, Ill.

Sydenham, E. W., H. F. Vismer, W. F. O. Marasas, N. I. Brown, M. Schlechter, and J. P. Rheeder, 1997. The influence of deck storage and initial processing on patulin levels in apple juice. *Food Addit. Contam.* **14**:429–434.

Taniwaki, M.H., J. I. Pitt, A. A. Teixeira, and B. T. Iamanaka. 2003. The source of ochratoxin A in Brazilian coffee and its formation in relation to processing methods. *Int. J. Food Microbiol.* **82**:173–179.

Thiel, P.G., W. F. O. Marasas, E. W. Sydenham, G. S. Shephard, and W. C. A. Gelderblom. 1992. The implications of naturally occurring levels of fumonisins in corn for human and animal health. *Mycopathologia* **117**:3–9.

Trigo-Stockli, D. M. 2002. Effect of processing on deoxynivalenol and other trichothecenes. *Adv. Exp. Med. Biol.* **504**:181–188.

U.S. Department of Agriculture (USDA), Agricultural Marketing Service. 2007. Privileged Communication.

Walker, R. 2002. Risk assessment of ochratoxin: current views of the European Scientific Committee on Food, the JECFA and the Codex Committee on Food Additives and Contaminants. *Adv. Exp. Med. Biol.* **504:**249–255.

Williams, J.H., T. D. Phillips, P. E. Jolly, J. K. Stiles, C. M. Jolly, and D. Aggaarwal. 2004. Human aflatoxicosis in developing countries: a review of toxicology, exposure, potential health consequences, and interventions. *Am. J. Clin. Nutr.* **80:**1106–1122.

Wood, G. E., and M. W. Trucksess. 1998. Regulatory control programs for mycotoxin-contaminated food, p. 459–481. *In* K. K.Sinha, and D. Bhatnagar (ed.), *Mycotoxins in Agriculture and Food Safety.* Marcel Dekker, New York, N. Y.

Wood, G. E., M. W. Trucksess, and S. H. Henry. 2003. Major fungal toxins of regulatory concern, p. 423–443. *In* M. D. Miliotis and J. W. Bier (ed.), *International Handbook of Foodborne Pathogens.* Marcel Dekker, New York, N. Y.

Zinedine, A., J. M. Soriano, C. Juan, B. Mojemmi, J. C. Molto, A. Bouklouze, Y. Cherrah, L. Idrissi, R. El Aouad, and J. Manes. 2007. Incidence of ochratoxin A in rice and dried fruits from Rabat and Sale area, Morocco. *Food Addit. Contam.* **24:**285–291.

Imported Foods: Microbiological Issues and Challenges
Edited by Michael P. Doyle and Marilyn C. Erickson
© 2008 ASM Press, Washington, DC

Role of Programs Designed To Improve the Microbiological Safety of Imported Food

Ewen Todd and Julie A. Caswell

8

INTRODUCTION

Globalization has created both benefits and concerns. The creation and trade of food commodities is a part of the pros and cons of globalization, and the future for cheap, available, and affordable food for all people is still on the horizon. The United States is increasingly relying on imported raw products and fully processed items to satisfy its consumer demands fueled by an affluent lifestyle, ready-made convenience products with the minimum of preparation, and awareness of new and exciting products through travel and through the media including the worldwide web. The North American Free Trade Agreement (NAFTA) between Canada, Mexico, and the United States has resulted in some challenges in the food and food safety arena. U.S. agribusiness challenges Canadian provincial and federal government restrictions on the export of milk, and Canadian Cattlemen for Fair Trade challenge the U.S. ban on imports of Canadian live cattle and beef following the discovery of a bovine spongiform encephalopathy (BSE)-infected cow from an Alberta herd (Sinclair, 2007). At the international level, the Doha Round of trade talks is no further ahead after 6 years and nothing more is expected until 2010 (Heller, 2007). The aim of the Doha Round, launched in November 2001, was to free global trade by cutting industrial and agricultural tariffs and by reducing farm subsidies, with specific focus on achieving concrete benefits for developing countries. The European Union (EU) food industry believes that a successful conclusion to the Round would help secure continued industry investment in

Ewen Todd, Director, Food Safety Policy Center, and Professor, Department of Advertising, Public Relations and Retailing, 328 Communications Arts and Sciences Building, Michigan State University, East Lansing, MI 48824. Julie A. Caswell, Professor and Chairperson, Dept. of Resource Economics, 215 Stockbridge Hall, University of Massachusetts, 80 Campus Center Way, Amherst, MA 01003-9246.

Europe, push trading partners into reforming their agricultural systems, and provide greater trading opportunities for food and drink manufacturers.

Another concern that will see global disorientation for the food supply is climate change. This will bring about challenges to the production of existing commodities as some subtropical agricultural areas, where many of the developing countries are becoming more arid and land in the higher latitudes is opening up for crops and livestock. Within a few decades, this may put a severe burden on world food production, with new trading structures and partners emerging. For instance, in a country already experiencing drought on a regular basis, the 2007 drought in Australia meant wheat yield reductions of up to 50%, and grain value was expected to undergo about a 21% increase in 2007 over the previous year (Eyre, 2007). In the past few years, changing weather conditions have threatened Australian stocks; genetically modified (GM) foods may be a technology that is necessary to counter the effects of global warming. Scientists are setting up trials for 30 GM wheat lines, each containing genes for drought tolerance derived from maize, thale cross, moss, or yeast. The GM debate in Australia is going to continue to be controversial, with many governments and wildlife organizations arguing the benefits and disadvantages of the use of GM food. Europe also should not consider itself immune from climate change, and Peter Mandelson, EU trade commissioner, argues that if it does not work through the issues raised by GM food it will not be working in its own interests and will open itself up to economic risks (Eyre, 2007). Thus, innovative approaches, such as development of drought- and heat-resistant GM products, will likely have to be taken to maintain seafood, dairy, crop, and meat animal supplies for the world's multiplying population. The United States, therefore, has to position itself, not only now but in the future, to anticipate how its growing demand for imported food will be met through supporting international research and political partnerships. At the same time, conflicting issues often emerge and must be resolved in order to implement an effective strategy to improve the microbiological safety of imported food, as illustrated in the two boxed examples that follow: (i) BSE and the risk of variant Creutzfeld-Jakob Disease (vCJD) from eating beef, and (ii) *Salmonella* in eggs and broilers. This chapter will describe the range of tools we have at our disposal that may be adopted to improve the microbiological safety of imported food.

RISK ANALYSIS

As discussed in Chapter 2, food safety in the United States is a shared responsibility, where several departments of the federal government share jurisdiction over ensuring the safety of the American food supply. In addition

BSE and the Risk of Variant Creutzfeld-Jakob Disease

The United States is concerned with both the import and export of cattle and beef products, in part because of the risk of bovine spongiform encephalopathies (BSE). The United Kingdom (UK) experienced BSE in older cattle in the mid-1980s which developed into the first mad cow disease outbreak and linked variant Creutzfeld-Jakob Disease (vCJD) causing human deaths from eating beef products containing nerve tissue. Imports of beef products from the UK were quickly stopped around the world, and approximately 4.5 million head of cattle were destroyed at a cost of $4 billion (FDA, 2003). Nearly 10,000 people in the transport and meat processing industries were laid off as a result of the mad cow crisis. However, the impact went further than the UK. When other countries experienced BSE in their herds, which were assumed to have been contracted through imported animal feed, global trade became further affected. Specified risk material (SRM), including any brain or nerve tissue, was removed as a cattle feed ingredient. In 1997, the United States and Canada instituted animal feed safeguards in an attempt to stop the spread of BSE, including a broader BSE testing program. However, the discovery of a mad cow in Canada in May 2003, and the subsequent discovery of a BSE-diseased cow, imported from Canada, in Washington State in December 2003 indicated these safeguards were insufficient and dramatically affected both the Canadian and American beef industries. After these discoveries, 53 countries, including such high-consumption markets as Japan and South Korea, forbade the importation of American beef (Hedges, 2007). Initially, the USDA had banned Canadian cattle from entering the United States. However, to aid the American meat industry suffering from loss of inputs, the USDA proposed lifting its ban on the importation of cattle from Canada, despite Canada's discovery of eight cases of BSE since 2003, including five in 2006 alone. The USDA risk assessment indicates that Canada's beef-raising and feeding practices have little risk for U.S. consumers to contract vCJD. Smaller American operations claim that Canadian imports will drive down the price of their own cattle and want to restore the ban. The larger companies, which control about 80% of the U.S. market, would benefit from a resumption of the Canadian imports by providing a source of cheaper live cattle and possibly the importation of processed beef. In 2005, the USDA allowed the renewed importation of younger Canadian cattle—those less than 30 months of age—since BSE is thought to occur only in older cattle. It is also possible that older Canadian cattle are being allowed across the U.S. border by USDA, as stated by Food and Water Watch (Gorham, 2007). This watchdog group is opposed to U.S. plans to resume trade in older cattle. Workers at slaughterhouses in three states have claimed to have had direct orders from supervisors not to intervene when older Canadian animals are being processed in violation of regulations. This is despite the fact that both Canada and the United States received a "controlled risk" designation, the second-highest safety label for beef safety from the World Organization for Animal Health (OIE) in 2007, and every animal coming from Canada for slaughter arrives with an import certificate that lists its age and identification in months.

Salmonella in Eggs and Broilers

In the UK, salmonellosis was associated with the consumption of sesame prawn toast in two outbreaks (Holtby et al., 2006). The causative agent, *Salmonella enterica* serotype Enteritidis phage type 14b, was isolated from many patients, from a packet of sesame prawn toast, and from imported Spanish eggs. These eggs had been used in the manufacture of the sesame prawn toast. The UK Health Protection Agency investigated more than 80 outbreaks of serotype Enteritidis (other than phage type 4) from 2002 to 2004, with at least 2,000 confirmed cases, and the evidence shows that the use by the UK catering trade of Spanish eggs is a major source of this infection (FSA, 2004). Since January 2004, these eggs have had to be marked "ES" (España), so that both caterers and consumers know that they will need to take extra care if they use these eggs, or they may choose to use UK eggs, marked "UK," which have had far fewer problems. This issue sharply resurfaced when the Food Standards Agency (FSA) found that of a total of 1,744 boxes of six eggs or more from European farms tested during March 2005 and July 2006, *Salmonella* contamination on the egg shell was found in 157 box samples (Reynolds, 2006; FSA, 2006). The FSA found that eggs from Spain had the highest rate of contamination, with an estimated one in every eight boxes testing *Salmonella*-positive. The Spanish Ministry of Agriculture, Fisheries and Food (MAPA) tried to discredit the FSA report published on 15 November, arguing that the report was based on samples taken from one packing center in Valladolid which packages eggs from only three farms out of a total of 1,100 farms that export eggs. However, the FSA stated that testing was conducted on stamped eggs in boxes ready for sale at the same time European Food Safety Authority (EFSA) was checking for, and finding, *Salmonella* at the same farms. Although *Salmonella* in imported eggs is not such an important issue for the FDA, this example illustrates the complexity of different public health and inspection systems in trading countries.

Variability in broiler contamination as with egg contamination is illustrated by the EU-baseline survey carried out in 2005 and 2006 to determine the prevalence of *Salmonella* in commercial flocks of broilers with at least 5,000 birds (EFSA Task Force, 2007). Average prevalence of *Salmonella*-positive flocks was 23.7%. This means that in the EU one in four broiler flocks raised over the one-year period of the baseline survey was *Salmonella*-positive, but even more disturbing was the variation in prevalence among the Member States, from 0% (Sweden) to 68.2% (Hungary). Other countries with high prevalence rates were Poland (58.2%), Portugal (43.5%), and Spain (41.2%). Thus, trade between Member States and with countries outside the EU has to take into account this variability in broiler contamination.

to strict regulations, the safety and wholesomeness of U.S. food products are safeguarded through premarket clearances, mandatory production practices, inspections, and random, ongoing sampling. Unfortunately, management of the food supply and oversight of responsible agencies and suppliers has been questioned by the General Accounting Office and watchdog organizations for many years (GAO, 2007; Food and Water Watch, 2007). As an illustration of potentially inefficient utilization of resources, both the USDA's Food

Safety and Inspection Service (FSIS) and the FDA used computer systems to review information on each import shipment to help identify the import shipments requiring inspector action (GAO, 1998). However, neither agency's system took maximum advantage of available data to target those imported foods posing the greater health risks. While the FSIS relies primarily on the violation history of previous shipments from the exporting firm to target entries for inspections or laboratory tests, the violation history may not always indicate the shipments more likely to pose health threats. For example, many violations, such as incorrect shipping labels, may not directly affect consumer safety, such as occurred in 1996, when ≈86% of FSIS-refused shipments, excluding those refused entry for transportation damage, were for reasons not directly related to health risks (i.e., excessive residues, microbiological contamination, unsound condition, or defects caused by disease) (GAO, 1998). Nevertheless, these violations triggered a series of inspections on subsequent shipments of the same product from the same exporting firm until at least 10 consecutive shipments were found to be in compliance. When limited resources are targeted in this fashion, fewer resources are available for products posing greater health risks.

Since the 1990s, the application of risk analysis principles has been an increasingly common focus in the development of regulatory and private assurance programs for both domestic and imported microbiological food safety. Risk analysis includes risk assessment, risk management, and risk communication in an iterative process that is intended to be updated over time as new information becomes available. In the application of risk analysis, consideration should be given to the following:

• evidence of actual or potential hazards to health
• microbiological status of the raw materials
• effect of processing on the microbiological status of the food
• likelihood and consequences of microbial contamination and/or growth during subsequent handling, storage, and use
• categories of consumers
• cost-benefit ratio associated with application of the criteria
• intended use of the food

The boxed example on assessment of imported seafood illustrates the application of the risk analysis approach.

CODEX ALIMENTARIUS

International organizations play important leadership roles in efforts to improve the microbiological safety of foods. Among these organizations, the Codex Alimentarius Commission (Codex) is particularly important.

Assessment of Imported Seafood

More than 80% of the U.S. seafood is imported (National Marine Fisheries Service [NMFS], 2007). Between 1990 and 2004 there were 984 seafood-related outbreaks (9,969 cases): 350 were caused by scombrotoxin (bacterial spoilage of finfish like tuna), 87 by *Vibrio* species (shellfish), 85 by norovirus (shellfish), and 60 by *Salmonella* (both finfish and shellfish) (Smith DeWaal et al., 2006). This is despite the fact that the FDA requires Hazard Analysis and Critical Control Point (HACCP) certification for seafood processors, which also applies to imported products. FDA rejects seafood for import mainly for being filthy (in whole or in part of a filthy, putrid, or decomposed substance, usually microbiological spoilage), and for *Salmonella* and *Listeria* in both cooked, ready-to-eat products and raw, frozen products (Huss et al., 2004). Sometimes histamine is detected in the fish, indicating potential scombrotoxin. Rejections also occur if there is no credible HACCP plan for producing the seafood. Predictive microbiology has been used to determine the risks of certain pathogens, typically *Listeria*, growing in certain seafood products. For *L. monocytogenes* in cold-smoked salmon and *Vibrio parahaemolyticus* and *V. vulnificus* in raw oysters, predictive models have been included in exposure assessment models for risk assessments (Ross et al., 2000; Buchanan et al., 2004). These types of models can be incorporated into risk assessments to help determine the levels of the pathogens at the time the products are consumed, and thus estimate the risk of illness. Several quantitative risk assessments have been done: *V. parahaemolyticus* in raw oysters harvested and consumed in Australia, Canada, Japan, New Zealand, and the United States; *V. parahaemolyticus* in finfish consumed raw; *V. parahaemolyticus* in bloody clams harvested and consumed in Thailand; and *V. vulnificus* in raw oysters harvested and consumed in the United States. The most relevant, however, is the assessment of choleragenic *Vibrio cholerae* O1 and O139 in warm-water shrimp in international trade (Bowers et al., 2006). This risk assessment was undertaken to use the available data to address some of the problems faced by developing countries with respect to the export market for warm-water shrimp. Outbreaks of cholera have been associated with consumption of seafood, including oysters, crabs, and shrimp (Oliver and Kaper, 2007), and most significantly a pandemic of cholera occurred in Latin America beginning in Peru in 1991. Some illnesses spread to the United States with crabs, raw oysters, and possibly ceviche (marinated raw fish) implicated as vehicles of transmission (Anonymous, 1991). The United States is the top importing nation for shrimp (all product types), with 345,700 tonnes entering the country in 2000. Consumption has steadily increased from 96×10^7 servings in 1996 to 125×10^7 servings in 2000, although relatively few people eat this product regularly (5%). In that same year, the FDA inspected shipments from India (148), Ecuador (16), Chile (14), Mexico (10), Thailand (5), and Venezuela (2), and none of these was positive. Four cholera cases were recorded for that year but not necessarily linked to shrimp. The risk characterization indicated that the median risk was between 2 and 9 illnesses for every 10^9 servings of warm-water shrimp. The median risk of acquiring cholera from warm-water shrimp in seven importing countries ranged from 0.009 to 0.9 cases per year depending on the country (Japan, United States, and five European countries). Predicted illnesses were very low, with one to two cases of cholera caused by consumption of warm-water shrimp in a decade predicted for Japan, the United States, and Spain.

Established in 1963, Codex is a joint effort of the United Nations Food and Agriculture Organization and the World Health Organization. Currently more than 160 countries from all parts of the world are members of Codex. It develops food standards, guidelines, and related texts such as codes of practice under the Joint FAO/WHO Food Standards Program. Its focus is on protecting the health of consumers and ensuring fair trade practices in the food trade, as well as promoting coordination of all food standards work undertaken by international governmental and nongovernmental organizations.

Codex gained added prominence in the 1990s under the Sanitary and Phytosanitary (SPS) Agreement of the WTO, which recognizes the desirability of harmonized standards between countries. Adoption of standards developed by Codex is encouraged in that, if a country's regulation adopts a Codex standard, the regulation cannot be challenged based on its risk assessment. Countries retain the right to choose a risk standard that is different from the Codex standard but must be able to support it with scientific risk assessment. This added prominence raises the stakes in Codex proceedings and may slow down the standards-setting process.

Codex carries on its work through a set of committees and task forces that develop standards and advice. The standards are developed by a multistep process and may have a commodity basis (e.g., evaporated milks, canned salmon) or be general across commodities (e.g., general methods of analysis for food additives). Codex has developed extensive standards, codes of practice, and guidance on microbiological food safety. As one example, the General Standard for Whey Cheeses notes under hygiene that these products should be "prepared and handled in accordance with the appropriate sections of the Recommended International Code of Practice—General Principles of Food Hygiene, the Code of Hygienic Practices for Milk and Milk Products, and other relevant Codex texts such as Codes of Hygienic Practices and Codes of Practice" and "should comply with any microbiological criteria established in accordance with the Principles for the Establishment and Application of Microbiological Criteria for Food" (CAC, 1997b).

The Codex standard-setting process and the standards themselves play an influential role in microbiological food safety for imported foods on several levels. On the most general level, over time Codex defines good regulatory practice, which is then applied in the regulation of domestic, imported, and exported foods around the world. It has done sustained work in defining good practices for risk analysis, including for risk assessment, management, and communication. Codex standards are a guide to regulatory standards set by countries for domestic, exported, and imported foods. In addition, the Codex Committee on Food Import and Export Inspection and Certification Systems develops standards that apply to the operation of trading systems.

Through its standard process and other capacity-building activities, Codex provides important support to the food safety programs of countries with limited resources for developing standards and enforcement programs.

Criticism of the Codex system, including a Codex-sponsored review (2002), has focused on problems with the timeliness and efficiency of the standard-setting process and on inadequate opportunities for developing countries and countries in economic transition to participate effectively in the process. Additional concerns are strengthening the scientific base for risk analysis and more effective capacity building for the development of national food control systems. A more fundamental challenge facing Codex in its role of promoting the microbiological safety of food products, including imported foods, is whether its public, collaborative, and iterative standard-setting process will increasingly be preempted by national standards, particularly of developed countries, and private standards that are established and evolve much more quickly than those of Codex. The interlaid impact of Codex standard setting is seen in the boxed example that follows.

AUSTRALIA/NEW ZEALAND RISK-BASED APPROACH

The Australian, and to some extent, the New Zealand position on the safety of imported foods has the same goal as that of the United States and most countries: facilitate trade but not allow contaminated products to enter the domestic market. In contrast to the United States, in recent years Australia and New Zealand, through the Australia New Zealand Food Authority (ANZFA, or its predecessor FSANZ), have adopted a risk-based management approach that strives to achieve, not zero risk, but what seems to be acceptable risk in these countries. ANZFA's authority is found in the provisions of the *Imported Food Control Act 1992*. It has the responsibility for categorizing foods according to assessed risk; the foods are then inspected by the Imported Foods Program (IFP), established in 1990. Within IFP, foods are classified as Risk or Surveillance category foods. Risk category foods are initially determined and periodically reviewed by ANZFA on the basis of scientific risk assessments and are determined to have a high or medium potential risk of contamination or other defects of food safety concern, at frequencies likely to be unacceptable from a consumer safety perspective, particularly for microbial contaminants. Risk foods are subject to the most intensive inspection frequency. All these Risk category foods are referred electronically by the Australian Customs Service for inspection decisions by the Australian Quarantine and Inspection Service (AQIS) (2002). Provisions are built into the AQIS system to recognize compliant producers and penalize those that have failures. Foods in the Risk category from a new overseas producer begin with all

Codex Alimentarius Commission Standard Setting

In microbiological food safety, the Codex Committee on Food Hygiene has been the main arena for discussions on relevant food standards, including microbiological risk assessment, but other Codex Committees (Fish and Fishery Products, Fresh Fruit and Vegetables, Meat Hygiene, Milk and Milk Products) have been instrumental in establishing international standards. The criteria depend not only on the microbiological agent of concern, but also the analytical methods used and the sampling plan (CAC, 1997b). In addition, the microbiological limits should be considered appropriate to the food at the specified point(s) of the food chain. The General Guidelines on Sampling give detailed instruction on how to set up sampling plans, including 2- and 3-class attribute plans, and for imported foods (CAC, 2004). In most Codex documents, however, specific standards are not suggested but if they are needed, they should comply with any microbiological criteria established in accordance with the Principles for the Establishment and Application of Microbiological Criteria for Foods (CAC, 1997b). In the case of indicator microorganisms, e.g., generic *Escherichia coli*, Enterobacteriaciae, and total viable counts (aerobic plate counts), the presence and/or concentration of these indicator organisms should reflect states or conditions that indicate process control or lack of process control. For infant foods, standards for *Salmonella*, coliforms, and mesophilic aerobic bacteria are documented in the Recommended International Code of Hygienic Practice for Foods for Infants and Children (CAC, 1979). Codex also has guidelines to provide a framework for the development of import and export inspection and consistent certification systems intended to assist countries in the application of requirements and the determination of equivalency, thereby protecting consumers and facilitating trade in foodstuffs (CAC, 1997a). Codex is currently trying to get agreement on standards for *Listeria monocytogenes* in ready-to-eat foods; these national standards are generally <1/g up to 100 CFU/g. Since foods at risk, e.g., soft cheeses, pâtés, deli meats, and smoked fish, are frequently shipped globally, it is important not to have trade blocked by differing criteria. More detailed discussion on international *Listeria* standards is found in Todd (2007).

consignments being inspected. Once an overseas producer establishes a record of reliability by having five consecutive shipments of a particular food cleared without problems, inspection intensity drops to one in four shipments selected randomly. If 20 inspections at this rate are clear, and importation follows a steady pattern, inspection intensity drops to 1 in 20 shipments, again selected randomly. Surveillance category foods are divided into two categories: Active and Random. Foods in the Active surveillance category are considered to pose a moderate public health risk, and for which more information is needed to make a definite categorization. Ten per cent of shipments of designated active surveillance foods, from every supplying country, are referred for inspection. Foods in the Random Surveillance category are considered to pose a low level of risk to health and safety, and inspections are carried out on 5% of all consignments of such foods inspected, selected at random. Of the

---

Cooked Crustacea

Cooked crustacea is categorized as a Risk category food by the Imported Foods Program and 100% of consignments are referred to Australian Quarantine and Inspection Service for inspection at a performance-based rate (FSANZ, 2003). Microbiological tests currently include a standard plate count, staphylococcal enterotoxin, and *Salmonella*. Compliance with these existing microbiological standards ensures that cooked crustacea has been produced and handled by using good hygienic practices and provides sufficient assurance of the protection of public health and safety. However, microbiological testing of imported cooked crustacea products between 1995 and 1999 resulted in a low failure rate, with < 2% of samples failing. In addition, the risk assessment work revealed that the public health risk posed by *L. monocytogenes* in cooked crustacea was very low and, based on that, it was concluded that a microbiological limit for this pathogen in cooked crustacea was not justified.

17,685 microbiological tests performed on Risk foods for the five-year period 1995 to 1999, there were 486 failures (2.7%). Yearly failure rates ranged from 2.3 to 4.0% (Bull et al., 2002). Failure rates [(failed tests/total tests) × 100] over the five-year period revealed that of all the Risk foods, smoked vacuum-packed fish had the highest failure rate at 8.6% for *Listeria* contamination. Other foods that tested positive for *Listeria* included soft cheeses, chicken, and mussels; however, failure rates were only 1.1%, 0%, and 0%, respectively.

Awareness of trends such as the increase in *Listeria* failures is useful to help improve the overall quality of food entering Australia. Sixteen food recalls were prompted by a failed test under the IFP from 1995 to 2002, which seems to be a relatively small number. One of these was a multistate outbreak in 2001 attributed to imported halva, manufactured from sesame seeds and contaminated with *Salmonella*. In response to this outbreak, this product was then elevated to Risk category for 3 months and 100% of imports were tested for *Salmonella* contamination, during which there were no failures. These negative results led the agency to subsequently reclassify halva and other sesame seed products to the Random surveillance category. Additional examples that illustrate the application of this risk-based approach to the development of policies and standards by FSANZ are presented in the shaded boxes about cooked crustacea and Roquefort cheese.

CHINA AND THE IMPORT DILEMMA

In assessing the potential value of risk-based programs to improving the microbiological safety of imported foods, it is also necessary to consider how economics may challenge their value. This dilemma was brought to

---

Roquefort Cheese Imported into Australia and New Zealand

In 2005, after a ban of >10 years, importation of Roquefort cheese into Australia was permitted (Mercer, 2005). The approval covers only the sale of Roquefort raw milk cheese made under specific conditions in France. Other blue mould cheeses, whether imported or domestically produced, are not covered because inspectors approved only these production practices after visiting the cheese-making facilities in southern France, with acceptable European Union oversight in place. A risk assessment indicated that the risks of *Listeria* and other pathogens being in the cheese were very low. FSANZ was satisfied that this raw milk cheese had a level of safety equivalent to cheeses made from heat-treated or pasteurized milk, according to the Act. There is precedent for Roquefort. In 1998, the Australian and New Zealand authorities amended the food safety code to allow the sale of Swiss cheeses Gruyère, Emmenthal, and Sbrinz, all of which are made using unpasteurized milk. In 2007, Roquefort was also allowed to enter New Zealand because the food safety criteria set by the European Community are similar to those in place for the manufacture of New Zealand cheeses (Anonymous, 2007). Both countries admit there is an increased risk of gastrointestinal illness from these high-demand gourmet foods and warning labels will be put on the packages. Under Australia's food-labeling laws, a statement will be required on Roquefort cheese to indicate that it has been manufactured from milk that is unpasteurized and sourced from sheep. Interestingly, the New Zealand importer said that "each consumer has to weigh up the risk when eating it, and trust their cheesemaker."

---

the public's awareness with the 2007 recall of pet food containing ingredients imported from China that contained melamine and cyanuric acid (Canadian Food Inspection Agency [CFIA], 2007). With this recall, it became increasingly clear that the United States depends on inexpensive food ingredients from China in order to offer the consumer low prices for food products. Large multinational food companies are now reviewing their food safety procedures and upgrading equipment to anticipate the presence of similar contaminants in imported foods and prevent any future consumer backlash against Chinese or other imports. Such measures are deemed necessary in light of the World Health Organization's report that stated that China's convoluted regulatory framework, loose laws, and lack of monitoring has led to several food safety issues. At least nine ministries are involved in food safety, as well as other agencies, none of which has authority over the others, and there is a lack of coordination among these ministries; suppliers are not properly certified in the food chain; and the system is not adequately monitored on a regular basis (Chung, 2007).

Results of FDA food import inspections reveal that China is consistently among the leading countries with the most food shipment rejections. Most of the rejections were for chemical contamination (e.g., chemicals such as

Sudan Red, antibiotics, and pesticides in food). In the first four months of 2007, FDA inspectors refused 298 food shipments from China (Weiss, 2007). By contrast, 56 shipments from Canada were rejected, even though Canada exports to the United States about $10 billion in FDA-regulated food and agricultural products compared with about $2 billion from China. Although China is subject to more inspections because of its poor record of shipment failures, those figures indicate that the rejection rate for foods imported from China, on a dollar-for-dollar basis, is more than 25 times that for Canada. FDA inspectors return these rejected shipments to the Chinese importers, but frequently they return to U.S. borders, making a second or third attempt at entry because the shippers know that inspections are low relative to the volume imported into the United States. Change will likely be difficult in large part because U.S. companies have become so dependent on many China-grown and -processed foods and food ingredients that tighter import restrictions may harm the U.S. economy. The commercial interest of the United States is strong and has an interest in allowing food imports to enter the United States as quickly and smoothly as possible. For some food products there is little choice because China has driven many competitors out of business with its exceptionally low prices.

China strongly contests complaints that it is exporting tainted food and other products to the United States, saying controls there are improving and U.S. inspectors have approved 99% of its shipments during the past three years (Cody, 2007). Moreover, China contends that most of the blocked shipments were unauthorized and did not go through China's export control system. Their defense also includes the contention that China's record on food exports to the United States is slightly better than that of U.S. food exports to China. As an example, Chinese authorities detected *Salmonella* and "toxins" in U.S. pork and chicken exported to China, and as a result, food imports from the two U.S. companies involved were barred indefinitely. Such issues will likely continue in the future and even be exacerbated among large countries with diverse populations and the demand for increasing trade.

RISK MANAGEMENT AND COMMUNICATION

Risk management is the identification and selection of appropriate food safety controls based on risk assessment and other factors that may be important to the risk management decision. Producers, processors, distributors, consumers, and governments carry out risk management on an on-going basis. The Sanitary and Phytosanitary (SPS) Measures Agreement of the WTO sets some principles for risk management by governments in the area of food safety.

First, the country should be able to clearly link its chosen level of protection and regulatory goals to its risk assessment. In turn, the risk management options (e.g., standards and inspection systems) chosen should be linked to the desired level of protection and be as least restrictive to trade as is possible.

Risk communication is multifaceted, including communication throughout the risk analysis process such as between risk assessors and managers and between interested parties (e.g., companies, consumers) and risk analysts. The iterative and interconnected nature of the risk management and communication processes is evident in many areas of microbiological food safety. For example, a risk management option for reducing food-borne illness associated with pathogens on fresh produce may be to better communicate the risks from cross-contamination in kitchen settings to food service personnel and consumers. On the other hand, risk communication, for example, information about outbreaks associated with fresh produce, puts pressure on governments and food supply chains to adopt risk management practices that address sources of those concerns.

Risk communication to participants in the supply chain and to consumers is an important means of improving the microbiological safety of food products. For example, communication about the identification of new pathogens or pathogen/food combinations allows parties to adopt new risk management strategies to counteract those risks. Concepts of risk communication to consumers are broadening to include a comprehensive presentation of benefits and risks associated with consumption of different food products so that consumers can tailor their choices to their own circumstances (Board on Global Health, 2006; National Academies, 2007). However, most countries are likely to implement this approach in the context of mandatory standards that set a regulatory floor for food safety.

Although there are benefits to risk communication, companies and sectors of the food industry can still be protective and secretive about their problems. Such a situation occurred with the recent U.S. spinach outbreak in 2006 when the industry as a whole was slow in responding to illnesses (Todd et al., 2007), although later there was widespread cooperation with government agencies and academia regarding control measures and research. Another incident where the food industry was ineffective in its communication to the public and regulatory agencies occurred with avian influenza in turkeys in the UK (see shaded box). This case demonstrates the ease with which a zoonotic disease can spread from one area to another in the absence of a prompt response. Clearly, good risk management and communication are crucial to improving the microbiological safety of foods. A number of widely adopted strategies are discussed below.

Avian Influenza (AI) in Turkeys

In the UK, the virulent H5N1 strain was confirmed in February 2007 at the Bernard Matthews farm in Suffolk, England, one of the largest European turkey poultry operations. The authorities believed that the AI strain had entered the plant through importation of infected poultry material from Hungary, which had a number of outbreaks in ducks and geese in 2006 and January 2007 (DEFRA, 2007). Almost 160,000 birds had to be depopulated and hundreds of workers were laid off after turkey meat sales decreased by 40% (ElAmin, 2007). Poor hygienic practices were found at its plant that may have led to the spread of the disease between holding sheds. Hungary put restrictions on the movement and sale of poultry outside the affected areas as mandated by the EU. However, the EU did not impose a trade ban on the whole country. Bernard Matthews has a processing plant in Hungary near the outbreak area and regularly transported meat back to Suffolk. The farm was criticized for reporting the infections several days after they began.

HAZARD ANALYSIS CRITICAL CONTROL POINTS APPROACH

The Hazard Analysis Critical Control Points (HACCP) approach is a prime example of the application of risk management principles in a systematic manner. Under mandatory HACCP, companies are expected to identify hazards and critical control points, and to establish management plans that keep those hazards within tolerances. The discipline imposed by the HACCP approach, whether mandatory or voluntary, can contribute to ongoing improvement in food safety. However, successful operation under HACCP requires great vigilance, particularly in identifying new hazards. This is a key concern with imported foods and food ingredients that are sourced from around the world and, in some cases, from countries with regulatory environments that are not able reliably to ensure compliance with food safety standards.

SANITARY AND PHYTOSANITARY (SPS) MEASURES

The WTO Agreement on the Application of Sanitary and Phytosanitary Measures (SPS Agreement) of April 1994 gives each WTO member the right to determine its own level of SPS protection. Countries impose different standards and regulations to handle the risks of pathogen contamination from processing and other stages of production (Matthews et al., 2003). Differences in risk assessment and adoption of different risk management strategies lead to regulations that can vary significantly across countries. This frequently gives rise to barriers to trade. Under the WTO, these barriers may be considered legitimate based on each country's risk situation and approach to consumer protection. However, country-level regulations that discriminate between domestic and imported product, for example, by setting higher standards for imported products, are illegitimate under the WTO. In practice, challenging the regulatory standards and practices of trading partners is costly and time consuming with the result that many differences in regulatory approaches persist over long periods of time. The following shaded boxes explore two examples of variable standards.

Research has attempted to assess the impact of different standards on trade, but quantification is difficult. Trade is definitely affected, such as an exporter choosing not to trade with a country having a zero-tolerance standard if it does not make economic sense. Temporary bans are more difficult

---

Variable Standards for *Salmonella* in Poultry

In the United States and Canada, poultry producers add chlorine to the carcass rinse bath to reduce *Salmonella* and *Campylobacter* levels in retail birds. Any additive but water is illegal in the EU. Thus, there is virtually no trade between the United States and Europe for poultry. There is no requirement for elimination of these pathogens in U.S. and Canadian poultry since *Salmonella* is widespread in the environment and testing will find it, depending on the sample size (whole-carcass rinse versus swabbing a portion of the skin) and the increased sensitivity of newer analytical methods. A few countries impose near-zero or zero tolerances for *Salmonella* contamination in imported meat products, especially poultry. However, in the United States, an attempt by the USDA to introduce a domestic *Salmonella* performance standard did not succeed. The Supreme Court beef case in 2000 determined that *Salmonella* tests do not necessarily evaluate the conditions of a meat processor's establishment, and they cannot serve as the basis for finding a plant's meat adulterated.

The costs of preparing poultry products to meet the zero-tolerance import standards of some countries are extra, to be added to the costs for the implementation of Hazard Analysis and Critical Control Point (HACCP) measures required of U.S. federally inspected meat and poultry processors and slaughterhouses (Matthews et al., 2003). There could also be an increased risk for recalls. Chile, the Czech Republic, El Salvador, Honduras, and Slovakia apply zero-tolerance standards for *Salmonella* for imported poultry products. The United States argues that these five countries have double standards because they do not have sufficiently good eradication and surveillance systems capable of achieving zero tolerance in domestic products, but expect imports to have no *Salmonella*. Double standards are seen as trade barriers and are subject to action under the WTO. U.S. poultry exports to Russia have been stopped from time to time creating substantial reductions in U.S. production when *Salmonella*-positive test results occur, preventing the issue of an export certification. Although the United States is the largest poultry producer in the world with little need for poultry imports, the United States restricts poultry meat from countries if there is a risk of poultry diseases like Newcastle Disease affecting domestic flocks; thus, only Canada, France, Israel, and the UK are permitted to export fresh, frozen, and chilled poultry to the United States.

Variable Standards for *Listeria monocytogenes* in Foods

The variable standard issue also applies to *L. monocytogenes* in ready-to-eat products, except here it is the United States that has the stricter standard compared with the EU and some non-EU countries. The U.S. government (FDA and USDA) states it is not in a position to set a regulatory tolerance partly because it cannot identify the end users of the products, some of whom would be high-risk individuals with lowered immune systems (Todd, 2007). However, the FAO/WHO risk assessment of *L. monocytogenes* in ready-to-eat foods (2004) indicates that nearly all cases of listeriosis result from the consumption of levels exceeding 100 cfu/g by several orders of magnitude of the pathogen, even in immunocompromised subpopulations. This assessment also states that the vast majority of listeriosis cases are associated with foods that do not meet current standards for *L. monocytogenes* in foods, whether the standard is "zero tolerance" (e.g., <0.04 cfu/g) or 100 cfu/g. As an example, the assessment predicts that when the level of *L. monocytogenes* was assumed not to exceed 1,000 cfu/g, the number of cases in the United States would be less than 26 per year versus the estimated 2,500 (FAO/WHO, 2004). Thus, the United States can prevent imports of foods that have a history of some *L. monocytogenes* being present and that have been involved in outbreaks in the past, e.g., Camembert and Brie cheeses made with unpasteurized milk. Other countries may perceive this as an inappropriate restriction. There is, however, justifiable concern about raw milk. An example is Hispanic soft cheeses being made locally in Latin American countries, notably Mexico, and brought illegally across the U.S.-Mexican border to meet the needs of Hispanics living in the United States (Kinde et al., 2007). In this situation there are risks of listeriosis to those eating the cheeses, which are typically prepared in unapproved facilities and do not meet the minimum of hygienic standards. The demand is sufficiently high that itinerant vendors risk fines to bring such cheeses into the United States on a regular basis. In February 2008, the FDA proposed a standard that would allow RTE foods that do not support the growth of *L. monocytogenes* to have up to 100 CFU/g of food (FDA, 2008).

---

to respond to because plants may gear up for large sales only to be stopped on occasion, and a glut of product remains on the domestic market to be sold at reduced prices. It is possible but not certain that world poultry trade would be higher if countries were to harmonize around the same standards, whether high or low.

Antibiotic residues present in foods can prevent their export, and the FDA monitors certain foods for these residues to see whether they would exceed the acceptable daily intake of 1.5 mg per person per day. Such low levels are assumed not to develop antibiotic resistance in the gut flora. Some antibiotics, however, such as chloramphenicol, are considered unsafe at any level because they may lead to cancer; these have been used in operations to treat diseases of animals, particularly in aquaculture. If these are detected in aquacultured products, such as shrimp typically raised in Asian countries, the FDA can prevent their import to the United States.

In summary, each WTO member country has the right to set standards as its government sees fit to determine its own level of SPS protection. These levels may be based on the science available in that country or copy what other countries have done. Zero tolerance is seen as the strictest level but is probably not achievable in most countries and may not provide the desired impact on disease reduction. If there are effective differences in standards between domestic and imported products, with importers having to show that their products are pathogen-free, these policies will likely be regarded as trade barriers that violate WTO rules.

GOOD AGRICULTURAL PRACTICES (GAPS)

To increase the value of farm products and expand markets for farmers, GAPs, codes of methods for appropriate agriculture production at the farm level, are promoted by many governments, exporters, producers, and retailers around the world. In recent years, GAPs also address food safety issues because of the demand by industrialized countries for fresh foods, such as fruits and vegetables, that can be consumed uncooked, with minimal risk of causing illness. Each GAP requirement comes with criteria to meet the safety standards set by the importer. In addition, even more recently there are niche markets for products that are grown, harvested, and marketed under environmentally friendly conditions (ISO14000 standards; Rainforest Alliance ECO-OK standards) and with support for the farm workers' welfare (such as ETI baseline; fair trade standards). These standards pose challenges for exporting developing countries, including (i) attaining financial resources to respond to the required hygienic standards, such as appropriate irrigation and pesticide spraying equipment, harvesting and processing equipment, and cold-chain storage and transportation; and (ii) building the capacity to educate farmers and processors about risky practices and how they can achieve GAP levels through on-site training. Many farmers find these criteria too difficult to achieve on their own and need external resources. In some cases, a completely upgraded infrastructure for whole farming communities is required. These resources may come through support by the national governments, international organizations such as FAO, charities, or exporting companies willing to invest in ventures with the local farmers, or some form of partnerships.

There is much focus on GAPs for correcting inappropriate pesticide and antibiotic use that can lead to unacceptable residues in the final products and subsequent rejection by importers. However, spraying fields with pesticides, herbicides, or irrigation water can also lead to contamination by pathogens. Biological control, including botanical pesticides, needs proper training to be as effective as the chemically produced pesticides, but the use of cover crops and crop rotation with appropriately timed spraying can keep the weeds and pests down to manageable levels and also not have the residues. Such control can also open the market door for organically grown products that are currently in high demand.

OTHER MANAGEMENT STRATEGIES

Prioritization of control strategies for imported foods can be developed based on surveillance data by identifying human cases that may have been consuming imported food. However, without an outbreak scenario these are difficult to detect. Other avenues for identifying contaminated product include (i) investigation of consumer complaints and (ii) monitoring programs that include microbiological testing of selected imported products based on past history and checking on GMP and HACCP plans accompanying the shipments. The integrated Danish *Salmonella* surveillance program information in 1999 developed a mathematical model for quantifying the contribution of each of the major animal-food sources to human salmonellosis (Hald et al., 2004). Imported foods, mainly shell eggs, were estimated to account for 11.8% of the cases. Other food sources, such as domestically produced pork, had only a minor impact, whereas 25% of the cases could not be associated with any source. This approach of quantifying the contribution of the various sources to human salmonellosis proved to be a valuable tool in risk management in Denmark and provides an example of how to integrate quantitative risk assessment and zoonotic disease surveillance for prioritizing control strategies for imported food.

Both government regulations and private interests in protecting markets and reputations lead companies to adopt new risk management measures. For example, Golan et al. (2004) studied the adoption of different risk management innovations and strategies by companies in the U.S. meat industry. In the case of hamburger patty production, Texas American Foodservice Corporation developed its Bacterial Pathogen Sampling and Testing Program as a process innovation combining a sampling protocol/management system and the application of a patented testing technology. The use of systematic feedback on quality to suppliers of meat grind for patty production and close in-plant controls allowed the company to supply foodservice operators who were imposing higher standards. In designing regulatory programs, governments must weigh the right mix of mandates, enforcement, and reliance on private markets that will result in efficient production of safe foods (Garcia et al., 2007).

With imported food products, countries face choices among risk management strategies that attempt to ensure safety at points of production and processing in exporting countries, screen imported products at the point of entry, or use some combination of those approaches. For example, the European regulatory approach for foods primarily follows the first approach, whereas the U.S. approach for products other than meats primarily follows the latter.

PRIVATE STANDARDS AND CERTIFICATION

Programs to improve the microbiological safety of imported foods are conducted by governments, companies, and consumers themselves through their food preparation practices. Producers, processors, distributors, retailers, and foodservice operators play key roles in producing and maintaining the safety of domestic and imported foods. With globalization comes the necessity of controlling the quality of products that are sourced from very diverse areas and supply chains across the world. Private parties (e.g., processors, retailers, and foodservice operators) have strong economic incentives to assure safety to the extent that poor quality can be traced to products they sell, and a food safety incident harms the reputation and sales of the company (Golan et al., 2004). This effect is particularly strong for manufacturers of branded food products, for retailers in conjunction with their own private label products, and for chain food service operators. In addition, in markets such as the United Kingdom, the use of private standards and certification can demonstrate due diligence, thus reducing the potential liability of a company in the event of a food safety incident (Hatanaka et al., 2005).

Increasingly, companies have turned toward the use of private standards for processing and for food safety and other quality attributes. These may complement government standards or in some cases be more stringent. They may be developed by the companies themselves or by third parties (e.g., trade associations, certification companies). Private standards may be enforced directly by the buyer or through third-party certification, where a third party does inspection and certification to the standards. The costs of certification are usually borne by the party seeking certification, although sharing of costs takes place in some systems. Other combinations of standard setting and certification are also in use. For example, in the United States organic standards are voluntary and set by the U.S. Department of Agriculture, while certification is done by private certifiers who are overseen by the government. The major players in developing private standards and certification vary in different parts of the world. For example, in Europe the major chain retailers are the dominant force in private certification. The boxed example that follows illustrates the operation of such standards among large retailers in the case of EurepGAP.

The proliferation of private standards and related certification raises important questions regarding their roles in ensuring the microbiological safety

EurepGAP Private Certification of Fruits and Vegetables

During the 1990s, large European retailers instituted a diverse range of standards related to Good Agricultural Practices (GAPs); these standards were not harmonized. In 2000 in response to problems associated with the proliferating standards, the Euro-Retailer Produce Working Group (Eurep), made up of the largest Northern European retailers, developed a common generic standard for good agricultural practices. While voluntary, this new standard, called EurepGAP, became required for all who wished to supply fruit and vegetables to Eurep-affiliated retailers.

Retailers focused on fresh fruits and vegetables for quality assurance because these products are closely linked to consumer perceptions about the freshness and quality of retail food operations. Problems in this area would be damaging to the reputations of retailers. In interviewing leading German and British retailers, Sterns et al. (2001) found that they pursued quality assurance systems for fruits and vegetables for multiple reasons. Due to market concentration, they have the market power to impose standards. They take seriously their gatekeeper role in tracking production and quality attributes for consumers that consumers themselves may not be able to track through inspection of the product at the retail store. Finally, standards offer these retailers a comparative advantage in logistics. The impact of the EurepGAP standard for fruits and vegetables has been felt throughout the fruit and vegetable production industries around the world. For example, a recent study of the Portuguese pear industry found that among farms that chose to sell to the United Kingdom, EurepGAP adoption was more likely among farms that were affiliated with larger producer organizations, more productive, complying with quality assurance systems such as a Protected Designation of Origin (PDO), and being farmed by full-time operators (Souza-Monteiro and Caswell, 2006). Increasingly, compliance with private standards is a requirement for access to markets in developed countries.

of imported food. In some settings, the private standards are de facto mandatory for accessing a market because major buyers will not purchase product that is not certified to their standards. If the private standard is more stringent than the governmental standard, then the governmental standard becomes a safety floor for imported products that has real effect only for product moving outside certified supply chains. The growth of private standards and certification has fueled a discussion of what has come to be called coregulation—the design of regulatory programs that take account of private quality assurance programs in order to achieve the best food safety outcomes given available regulatory resources (Garcia et al., 2007). Critics may view some forms of coregulation with skepticism, fearing that that the regulatory process will be co-opted by the regulated firms. Certainly risk-based systems should take into account the stringency and reliability of the private standards applied to imported foods in making inspection decisions.

The growth in and proliferation of private standards, as well as the increasing stringency of governmental standards, have also raised questions of access to international markets, particularly for producers and companies operating in countries with less developed systems of food safety control (World Bank, 2005; Jaffee and Henson, 2004). For small producers, the costs of complying, which may include purchasing new equipment, redesigning buildings, installing contamination detection systems, and instituting laboratory testing, may be prohibitively high unless they receive subsidies or form cooperatives. In general, private certification may raise the short-run costs of suppliers but they may be compensated through higher prices or longer-term productivity.

Private standards and certification are established by buyers and complied with by sellers when the individual companies perceive their benefits from these systems to be larger than their costs. Harmonization of standards, or the establishment of meta-standards such as the ISO series, occurs when it is beneficial to supply chain partners that are in a position to make decisions for the chain. In a case such as the establishment of the EurepGAP system, large retailers banded together to set a common standard. Private standards and certification system will continue to evolve quickly, sometimes in tandem with government regulations, sometimes leading them, and at other times possibly confounding them.

COUNTRY-OF-ORIGIN LABELING

Country-of-Origin Labeling (COOL) may be mandated by countries in order to require the communication of the origin of food products along the supply chain (e.g., from producers to distributors to retailers) or to the final consumer. It can also be voluntarily adopted by producers, distributors, or retailers to differentiate their products from those that have their origins in other countries.

Whether COOL is a risk communication tool that can be designed to improve the microbiological safety of imported foods, or whether it does in fact do so, is open to debate. Two key links are necessary for the connection to be made between COOL and improvement in microbiological food safety. First, food safety must be reliably associated with products from particular countries so that a label communicates about the relative safety of a product. For example, consumers would need to know that product X from country Y is safer than the same product from country Z and buy the product from country Y more often or exclusively. Second, this type of buying signal would need to be transferred back down to actors in the supply chain or to governments, who would then respond by taking actions to improve food safety.

Thus, COOL may act as an incentive system but is not a direct food safety measure.

At the same time, COOL may be used to signal a range of quality attributes other than safety, including appearance, flavor, and color, again to the extent that these attributes are associated with food products originating in particular countries. Origin itself may in some cases be a quality attribute if buyers prefer foods from specific countries independent of the specific characteristics associated with the product (e.g., the consumer wants to support country Y and therefore buys product from country Y).

COOL is mandated quite widely by countries around the world. For example, in 2003, the U.S. General Accounting Office surveyed U.S. agricultural attaches in 57 countries and found that 48 required COOL for one or more of the following products it was studying: beef, pork, lamb, fish and shellfish, fruits and vegetables, and peanuts (GAO, 2003). Mandatory COOL programs raise significant trade questions (Josling et al., 2004) as they may act as a barrier to trade for all products originating in countries with poor reputations for quality regardless of the individual quality controls used by particular producers and associated supply chains. In the WTO context, COOL may be considered to fall under the SPS Agreement or the TBT Agreement, depending on the regulatory goal of the labeling. There is no blanket guidance on whether COOL is a legitimate regulatory program under the WTO, although challenges have been limited.

In some cases, COOL could conceivably even discourage food safety improvements in that individual producers in countries with poor reputations may not believe it is possible to secure export markets through improving safety. Alternatively, origin labeling may offer opportunities to enhance trade volume for countries with strong reputations for quality. A key question in both cases remains whether and to what extent buyers (in the supply chain or consumers) associate country of origin with safety levels.

Some studies of consumer demand show that consumers are willing to pay a higher price for a country-of-origin label, frequently because they believe that COOL information signals both product safety and quality (Caswell and Joseph, 2008). The label can allow the consumer to distinguish between domestic and imported products. Unterschultz et al. (1998) found that if consumers prefer domestic products and think they are safer, they are willing to pay for COOL information. However, overview analysis (Ehmke, 2006) of willingness to pay studies suggests that inconsistencies in study design make it difficult to draw conclusions about links between consumer beliefs or knowledge about the relative safety of products from different countries and demand for COOL. For example, studies may ask consumers about their valuation of multiple product attributes (including COOL), making it difficult to sort out which attributes are most important to consumers and why. Lusk et al. (2006) explore these issues in further detail, concluding that there are several consumer motivations for origin labeling and that recent work by economists has failed to adequately identify why consumers desire COOL.

Further issues affect the use of COOL as a means of improving the microbiological safety of imported food. As noted earlier, the safety of a food product may or may not be well correlated with its country of origin. This becomes further complicated with products that go through different stages of production, which may occur in different countries. An example is beef, in which animals may be born in one country and fed up to the time of slaughter in another country. Such supply chains can result in complex origin labels, depending on country level regulations, that do not effectively convey quality information about the final product.

When issues arise regarding the safety of imported foods, governments are likely to take more direct action to protect consumers than relying on country-of-origin labeling. For example, in the United States fresh fish and shellfish are subject to mandatory COOL. When testing in 2007 revealed problems with drug residues in some products imported from China, the U.S. Food and Drug Administration (FDA) imposed automatic detention at ports of entry for five types of farm-raised fish and shrimp (FDA, 2007b). The FDA also set out conditions under which an exporter can be exempted from detention based on providing specific information to the agency that demonstrates it has implemented steps to ensure its products do not contain the substances of concern and that preventive controls are in place. Thus, while COOL may in some cases be operative to signal degrees of safety above the regulatory standard, the standard itself and other private quality standards are likely to have the controlling effect over the microbiological safety of imported products.

Benefits of COOL may also be circumvented through illegal smuggling and mislabeling of products. For example, smuggling of meat, animal products, and live animals has always been and continues to be an important issue (Anonymous, 2006), as illustrated in the following four examples from developed countries: (i) African Swine Fever in Belgium in 1985 was associated with infected meat being illegally brought from Spain and fed to a boar; (ii) smuggling of meat into the United Kingdom has been suggested as the origin of the large and very public foot and mouth disease outbreak in 2001; (iii) H5N1 virus was found in Japan in a shipment of frozen duck meat from Shandong province of China in 2003; and (iv) in 2005, Taiwan destroyed 82 tons of animal products as well as 5,146 animals, including 5,070 birds smuggled from abroad. Recent smuggling events that have occurred in the United States are described in the shaded box that follows.

Smuggling of Imported Food to the United States

In 2006, USDA teams seized hundreds of thousands of pounds of prohibited poultry products from China and other Asian countries. Some were shipped in crates labeled "dried lily flower," "prune slices," and "vegetables," according to news reports (Weiss, 2007). It is unclear how much of the illegal meat slipped in undetected. This information is in agreement with China's statement that the refused shipments were mostly illegal (Cody, 2007). In another example, Chinese restaurants and Asian grocery stores in Michigan were searched in July 2006 for frozen poultry products smuggled from China in violation of an import ban sparked by avian influenza (CIDRAP News, 2006). The USDA had seized 1,940 pounds of illegal poultry believed to be from China from a warehouse in Troy, Michigan. The smuggled frozen poultry was packed in unmarked boxes or in boxes labeled as tilapia fish. The products included geese, ducks, and chickens that had intestines intact. Intensified official veterinary controls on warehouses supplying Chinese retailers resulted in the seizure of smuggled poultry products in Italy, and low pathogenic avian influenza strains (LPAI) were found in lung and trachea from both frozen duck and chicken carcasses (Beato et al., 2006).

Overall, country-of-origin labeling is likely to be a blunt instrument for affecting the microbiological safety of imported foods. Country level regulations influence the safety of products exported from particular countries and COOL allows buyers to select products from those countries. However, not all producers from countries with stringent regulations meet those standards and supply chains with strict private standards can produce safe product in countries with weaker regulatory controls. Thus, while country-of-origin labeling can communicate about quality attributes that are important to buyers, it does not always provide the close connection between product demand and signals to producers to improve safety that are necessary to lead to improvements in the microbiological safety of imported food products.

FOOD DEFENSE AND BIOTERRORISM

Food bioterrorism is an act or threat of deliberate contamination of food for human consumption with biological agents for the purpose of causing injury or death to populations, as well as disrupting the economic and political stability of a country. The deliberate release of infectious pathogenic microorganisms or their toxins could place a long-lasting severe burden on public health systems. The globalization of the food supply means that such an event, or series of similar events, would likely have consequences beyond a single country, as evidenced by the many examples given in Chapter 3 of naturally contaminated foods from one country affecting another. Today, the media would report any deliberate attack and the news would quickly spread around the globe so that public fear and governmental and industry concern would be expressed. If these events were not contained immediately because of an inability to adequately respond, the production and shipment of foods around the world would be severely restricted by trade barriers being put up at the national and regional levels. Thus, there is a need for a new form of control—food defense that builds on food safety but is not identical to it.

In the United States in the wake of the September 11 attacks, the *Bacillus anthracis* contamination of a few letters raised the likelihood that an unknown number of terrorist events could occur even if few illnesses and deaths occurred, and these caused considerable disruption and public anxiety. Former Department of Health and Human Services (DHHS) Secretary Tommy Thompson publicized the risk of deliberate contamination of food in December 2004 by saying he didn't know why an attack had not already occurred, and raised both concern and controversy in his comments (Board on Global Health, 2006). If something like this does occur, it would be very unlikely to affect the entire food supply because of its diversity but would be very disruptive to the economy, as evidenced by the cyanide scare that occurred with Chilean grapes in 1989 (see shaded box). Severe restrictions of

---

Cyanide Scare in Chilean Grapes

Contamination of Chilean grapes with cyanide in 1989 led to the recall of all Chilean fruit from Canada and the United States, and the publicity surrounding this incident resulted in a boycott by consumers. The resulting damage amounted to millions of dollars and more than 100 growers and shippers who went bankrupt (Anonymous, 1993). The scare began when the American Embassy in Santiago received two telephone calls in early March 1989 warning that Chilean fruit shipped to the United States had been poisoned. This warning led FDA investigators to tighten inspections of Chilean produce entering the United States, and on 12 March 1989, the agency said it had found cyanide in two seedless red grapes taken from a ship docked in Philadelphia. The next day, the agency urged consumers not to buy Chilean fruit and ordered dock workers not to unload Chilean fruit aboard ships. The ban was lifted four days later after another test on the same two grapes failed to detect cyanide, but some fruit had spoiled, and sales of Chilean fruit around the world were affected. However, a federal judge ruled in 1993 that the FDA was not responsible for $210 million lost as claimed in a suit by 2,500 Chilean fruit growers and exporters and several American importers. The 2007 contamination of pet food with melamine for protein enhancement described elsewhere in this chapter is also likely to have multimillion dollar consequences to U.S., Canadian, and Chinese companies. Thus, analysis of the importing process is critical to limiting opportunities for deliberate contamination whether as a terrorist attack, personal revenge by a disgruntled employee or competitor, or for profit using illegal additives.

foods could immediately occur everywhere, and there would be food short-ages and reduced choices, with the public losing confidence in the govern-ment's ability to protect consumers. Deliberately contaminated imported food would certainly affect trade worldwide and probably erode the public's confidence in the government and other nations even further.

Potential modes for deliberate contamination of imported food are nu-merous. A terrorist could choose to use a rapidly acting severe and heat-resistant toxin such as staphylococcal enterotoxin or possibly a seafood toxin like paralytic shellfish poison (e.g., saxitoxin). Both would still be toxigenic after normal cooking. A terrorist could even use the more deadly *Clostridium botulinum* heat-labile toxin for effect within several hours. In these cases, the incriminated food would be quickly found and taken off the market. Or there could be a much longer incubation period of days to weeks with an infectious agent that would not be detected until most of the food had been distributed and eaten. There are many difficulties for the would-be perpetrators to carry either of these scenarios off, especially with imported food, but this is by no means impossible. For example, many developing countries providing foods and food ingredients to the United States lack basic food safety infrastructures and are vulnerable to deliberate acts of sabotage. Moreover, many foods are produced in centralized operations and distributed over large geographical areas, especially by multinational corporations. Even though foodservice op-erations have been the target of most criminal attacks, it is less likely that these would involve imported food unless the ingredients are contaminated. Contamination of ingredients or equipment at centralized facilities could af-fect large quantities of product that would be widely distributed. This is par-ticularly a concern when imported products are used as ingredients (e.g., spices, dried products, nuts) in other foods, which may provide an environ-ment more suitable for bacterial growth. Another source of contamination is at the transport stage. International and domestic transport of food is being handled increasingly in large containers and tanks. Tamper-resistant or tamper-evident locks or seals on containers should be used to detect any ob-vious attempts at tampering. Temperature controls and monitoring devices on refrigerated containers could be constructed to prevent unauthorized access. However, each container cannot be inspected and tested before delivery. How vulnerable tankers can be is illustrated by one of the largest *Salmonella* out-breaks in the United States. In 1994, an estimated more than 224,000 people in 41 states became ill after consuming ice cream (Hennesy et al., 1996). The pasteurized liquid ice cream premix had been transported in tanker trucks, one of which had a residue of liquid egg with serotype Enteritidis even after cleaning. The level of *Salmonella* contamination in this residue was sufficient to contaminate thousands of ice cream packages.

Where a deliberate contamination of imported (or domestic) food is a potential issue, the approach has to be different from investigating food-borne disease outbreaks and for unintentional contamination investigations. Investigators need to think about how perpetrators would have access to food and how an agent may be introduced into the vehicle. The delivery mechanism needs to be considered, including ways that the agent can be mixed and distributed through a food, the location of this vehicle, and what its potential is for spread into the environment. Expert opinion should be sought on the practical availability of the amount of agent required to contaminate a particular food item and have any adverse health impact (i.e., the amount required for a toxic or infectious dose with or without pathogen growth and multiplication).

According to the World Health Organization (WHO, 2002), the role of national governments is to take the lead in increasing the capacity of the national system to respond to food bioterrorist threats. The components of general preparedness plans that enable an effective emergency response include surveillance systems to detect a public health incident; assessment of vulnerability to the specific threat or incident; implementation of preparedness planning principles; testing preparedness plans for effectiveness; and capacities for investigation and verification of the threat or incident and linkage of the relevant government agencies and other bodies that will contribute to management of the public health consequences.

SURVEILLANCE SYSTEMS

Rapid identification of a food-borne disease agent and its source leading to the localization of the contaminated food product and its removal would reduce the number of illnesses and deaths. Many existing surveillance systems may have the capacity to detect clusters of food-borne disease, provided the cluster is large enough and the effects severe enough to cause people to seek medical attention. However, the capacity to do so depends on both recognizing the circumstances of the contamination and having sensitive enough investigative procedures. Rapid response depends on effective links to laboratories with the capacity to identify various food-borne agents, including unusual ones. Existing passive surveillance systems, however, are unlikely to provide early detection of the first cases of a large event, because only a small fraction of ill people seek medical care or submit samples for laboratory analysis, and laboratories typically only test for a few of the most common disease-causing agents to be reported to selected health officials. The anthrax-in-letters attack demonstrated that early detection and rapid response is not easy to do because only a few cases were scattered across different parts of the country with only the distinctive pattern of illness to link them at first (Tan et al., 2002). It is possible that less severe illnesses may have gone unrecognized. A sentinel surveillance system to rapidly detect small clusters of illness as the first signs of a potentially large food terrorism event needs to be set up to recognize symptoms that may be unusual and probably severe, and to be able to mobilize response teams to contain the contaminated products and remove them from potential consumers. In some situations, the first signs of a deliberate attack could be sick or dying animals, as occurred with the accidental poisoning of cows eating feed contaminated with polybrominated biphenyls (PBB) in Michigan in 1970 (Aust et al., 1987) and dioxin in animal feed in Belgium in 1999 (van Larebeke et al., 2001). Syndromic surveillance is one approach where improved reporting could result from better linkages with hospital medical records, pharmacies (for over-the-counter medication purchases), and poison information center reports. These could provide information about a potential outbreak in its early stages, and prompt investigation of an undetected disease cluster.

ASSESSMENT OF VULNERABILITY

A vulnerability assessment will attempt to identify the properties and potential consequences of deliberate contamination of food by harmful agents, to identify relative priorities, and to commit national resources in a proportion consistent with these priorities. The approach to such an assessment should be multidisciplinary with input from legal, intelligence, medical, scientific, economic, and political contributors. The following aspects should be considered even if minimal data are available: (i) the public health impact, on the basis of illness and death, of deliberate exposure to the agent (short- and long-term health effects, severity, vulnerable populations); (ii) the potential for delivery of the agent to large populations on the basis of its characteristics; (iii) the possibility of mass production and distribution of the agent; (iv) the potential for person-to-person transmission; (v) public perception, as related to fear and potential civil disruption; (vi) special needs for public health preparedness, including stockpile requirements; (vii) the need for enhanced surveillance or diagnosis; (viii) the potential of obtaining the necessary quantities of an agent sufficient to do harm; (ix) the identification of potential terrorists who would be motivated to carry out such an attack; and (x) the availability and effectiveness of preparedness plans and the capacity for effective response with general preparedness measures to contain the event and prevent further harm (WHO, 2002).

One approach to a vulnerability assessment that is widely accepted today in U.S. government agencies is the CARVER + Shock method, which is a tool designed for the military to prioritize offensive targets but has now been adapted for the food sector. It allows the user to assess and determine the most vulnerable points in a system or infrastructure, much like an attacker would.

The user can then focus resources on protecting these points. CARVER is an acronym for the attributes used to evaluate the attractiveness of a target for attack:

- *Criticality*—impact of an attack on public health and the economy
- *Accessibility*—ability to physically enter and leave a target undetected
- *Recuperability*—food system's ability to recover from attack
- *Vulnerability*—ease of attacking a target
- *Effect*—amount of direct loss to a target from the attack, measured by loss of productivity
- *Recognizability*—ease of identifying the target

A seventh attribute in this tool is the shock component to assess the public health, economic, and psychological impact of an attack (the "shock" value of a target) such as the difference in emotional shock between an attack on an infant formula plant, a fish-processing operation, and a company providing food for immunocompromised persons.

In the CARVER approach, each point in a food-processing system is ranked for each attribute according to its attractiveness on a scale from 1 to 10. Rankings utilize information collected from the food industry about the quantities of food produced, sources of raw materials, and distribution patterns and tracing systems for finished products. Conditions associated with low attractiveness or low vulnerability are assigned low values, while conditions associated with high attractiveness or high vulnerability are assigned high values. This information is necessary for estimating the scale of potential exposure and for removing the affected food from sale. The total is summed across the 10 attributes. Scoring the various elements of a food sector's infrastructure for each of the CARVER + Shock attributes helps identify where an attack on that infrastructure is most likely to occur. CARVER + Shock software, available from the FDA (FDA, 2007a), mimics the thought processes in play during a face-to-face CARVER + Shock session by having the user (i) build a process flow diagram for the system to be evaluated and (ii) answer a series of questions for each of the seven CARVER + Shock attributes for each process flow diagram node. Again, each question has an associated score. Based on the answers given, the software calculates a score for each CARVER + Shock attribute and sums them to produce a total score for each node.

IMPLEMENTATION OF PREPAREDNESS PROGRAMS

Government is responsible for introducing or strengthening existing emergency response systems to respond to food bioterrorism by identifying necessary components of an emergency response program. The Public Health Security and Bioterrorism Preparedness and Response Act of 2002 (Bioterrorism Act) Section 301 (Food Safety and Security Strategy) requires the President's Council on Food Safety, in consultation with the Secretaries of Transportation and Treasury; other relevant federal agencies; food industry, consumer, and producer groups; scientific organizations; and the states to develop a crisis communication and education strategy regarding bioterrorist threats to the food supply, including those from imported food (FDA, 2002). The strategy should be broad enough to address threat assessments, technologies and procedures for securing food processing and manufacturing facilities and modes of transportation, response and notification procedures, and risk communications to the public. The programs should be built on community, state, and national civil defense systems for responding to crises and emergencies. Here coordination among the agencies is critical. Plans should include scope of the plans and any assumptions made, responsibilities and organizations including relationships to law enforcement and emergency management, operating procedures including information flow, risk assessment/situation analysis, activating resources, deactivation, and recovery (IAFP, 2007). Individual roles should be clearly made, and emergency contact information for after hours should be kept current. It is important to establish which are the appropriate legal authorities and procedures for embargo, detention, and seizure of foods suspected to be intentionally contaminated. In connection with this is the establishment of sampling protocols with a chain of custody procedure for these foods.

The Bioterrorism Act affects importers of food products to U.S. ports (FDA, 2002). It can give high priority to increasing the number of inspections of food offered for import, with the greatest priority given to inspections to detect intentional adulteration. As explained in Chapter 2, the Act requires, with limited exception, the registration with the FDA of any facility that processes, packs, manufactures, or holds food to be consumed in the United States. In addition to registration, the Act also requires that these facilities maintain detailed records of the acquisition, production, distribution, and sale of food products; that registrants are immediately accessible to the FDA; and that registrants are able to provide the FDA with detailed tracing of any food product at short notice. In addition, foreign processors must designate a United States-based agent to serve as a contact point for the FDA. The Act requires that FDA receive prior notice for food imported or offered for import into the United States. Since there are estimated to be more than 9 million notice submissions every year, it is not feasible for checks to be made on all the consignments. Although this Act is supposed to protect the U.S. food supply from deliberate contamination by a terrorist, it is not clear how effective this has been. One major unintended consequence is an enhanced degree of control by the foreign manufacturer on pricing. Under the Act, an importer can only import product into the United States if it has the manufacturer's bioterrorism identification number. Due to confidentiality of the number, this number can only be provided directly by the manufacturer or by someone to whom the manufacturer has provided the number. By keeping the number confidential, the manufacturer can better control who sells its product to the United States.

TESTING PREPAREDNESS PLANS

Once a plan is set up, working relationships with the appropriate agencies and industry and other stakeholders should be set up; they should be in regular contact and have a command center set up for any role-playing exercises. Public health and criminal investigations are different, and roles and actions need to be clearly spelled out before any emergency occurs. Identifying gaps in the existing or proposed response system can be checked through regular communication, face-to-face meetings, and conference calls. The effectiveness and completeness of the emergency response system can be tested by simulated exercises or case studies. These can be started within an organization and then in combination with other agencies within the response plan. Such exercises include so-called table-top exercises where all potential players in detecting and containing a real attack would sit at tables and work out strategies as the scenarios unfold through a facilitator. A full-scale exercise is both time-consuming and expensive to carry out but is the most realistic way to assess the effectiveness of response teams.

CONFIDENTIALITY ISSUES

One of the issues raised with bioterrorism is publication of research into deliberate attacks and whether essential information could be given to would-be perpetrators (Alberts, 2005). Wein and Liu (2005) published an article in the Proceedings of the National Academy of Sciences on the deliberate contamination of a holding tank at a dairy farm, a tanker truck transporting milk from a farm to the processing plant, or a raw milk silo at the processing facility with *C. botulinum* toxin. They stated that "<1 g of toxin is required to cause 100,000 mean casualties (i.e., poisoned individuals), and 10 g poison the great majority of the 568,000 consumers" who consume the contaminated batch of milk. Most of the casualties would occur on days 3 to 6 from production. This public release of data raised many questions, and it was pointed out that these numbers may be an overestimate since protection of milk and other food facilities had improved since the 2001 anthrax attacks

(Alberts, 2005). In addition, a system of self-governance by scientists was recommended in which scientific journals are to apply special scrutiny to publications that:

1. would demonstrate how to render a vaccine ineffective
2. would confer resistance to therapeutically useful antibiotics or antiviral agents
3. would enhance the virulence of a pathogen or render a nonpathogen virulent
4. would increase transmissibility of a pathogen
5. would alter the host range of a pathogen
6. would enable the evasion of diagnostic/detection modalities or
7. would enable the weaponization of a biological agent or toxin.

However, it was also pointed out that all of the critical information in this article that could be useful to a terrorist—in particular, the LD_{50} of botulinum toxin for humans, toxin heat sensitivity, milk pasteurization conditions, and the size of the milk containers into which milk collections are pooled for pasteurization are accessible on the World Wide Web through a simple Google search (Alberts, 2005).

TRACEABILITY

The Principles

Traceability is the ability of a company, a retail chain, or an industry sector to trace the history of a product from production and distribution to consumption or use in another product. The need for traceability comes from two directions: logistics and safety. Traceability is a key component for reducing risks of microbial and chemical contamination by reviewing the production, harvesting, and production practices in the information recorded through the tracking system. Also, if a deliberate attack were planned, limiting its effect or spread should be facilitated by rapidly identifying the source through such a system. An increasing demand by consumers for safe and high-quality food has led governments, agricultural production operations, and food industries to explore ways of maintaining control of products (Dagg et al., 2006). End-product testing for pathogens or contaminants is recognized today as not adequate to ensure the safety of the final product (Butler et al., 2003). This is the case because of the relative rarity of these in the vast quantities of food being produced domestically and overseas, as well as the delays involved in testing and holding. However, the expected standards are more likely to be met if an importer can understand and somewhat control the whole production process. For instance, the use of quality assurance systems, based on sound risk management principles, including GAPs, Good Manufacturing Practices (GMPs), HACCP, and time-temperature monitoring of transportation, means that the imported food has a more consistent quality and is less likely to have contaminants, at least in unacceptable numbers.

Adoption of a traceability system by agriculture and the food industry should complement existing government policies. Typically, traceability systems facilitate control through traceforward (allowing the withdrawal of a product) or traceback (to determine its source) at each stage of the food supply chain. If there are no problems, some kind of certification or approval step is given to proceed to the next stage. The approval part can be determined directly through negotiation between the purchaser and the supplier, but often third-party contractors ensure that the quality assurance systems are implemented and verified to be in compliance with domestic requirements, as well as those of the importing country (government and industry).

An effective tracking system for imported food can capture information from data loggers related to these control mechanisms. Such information can be downloaded at an inspection location in the importing country. However, if the data indicate a degree of dissatisfaction with a risk to quality and safety, the only option for the importer is to return, destroy, or reprocess the product. If the monitoring can be done throughout the food chain in the exporting country, unsatisfactory readings at earlier stages may mean the stages in question can be checked out for lack of compliance with company guidelines, and the necessary controls introduced where appropriate. The product itself may either be returned to the manufacturer locally and avoid costly recalls in the importing country or be tested for contaminants and other measures of insanitation and, if satisfactory, be allowed to proceed. Traceability monitoring systems are becoming more sophisticated through advances in technology that allow more and more information to be recorded in microchips. However, even paper trails can be useful for tracking if properly used throughout the whole food production and exporting chain. If tracking systems for safety and quality are also tied in with economic incentives, there is a much better likelihood these will be adopted. For instance, payment may be made at different stages in the shipment from the production site to final destination and these payments may be held up until the shipment has satisfied the importer's conditions in order to proceed.

The main motivation for companies in food chains to consider traceability systems is the concern for food safety. Adoption of traceability allows a company not only to retrieve a contaminated lot quickly and not implicate similar lots or even brands with loss of consumer confidence, but also to more easily assign blame to the part of the chain where the problem occurred

(Verdenius, 2006). Nevertheless, even though safety issues are also more easily controlled though appropriate traceability, the food industry was slower to recognize this than other industry sectors. As food distribution systems became longer and more prone to errors and larger quantities of food were recalled or destroyed in the 1990s, the industry as a whole began exploring traceability schemes with identifiable key nodes. The UK Food Standards Agency also identified several reasons for adopting traceability systems in the food supply chain: (i) to respond quickly to a food safety problem, e.g., foodborne outbreak or contamination, to protect the public through recalls and investigations; (ii) to facilitate food sampling at appropriate points in the food chain for pesticide and other residues and where excessive levels may have occurred; (iii) to comply with existing legislation; (iv) to provide quality assurance in food processing and handling; (v) to facilitate access to information to help in determining risk assessments; (vi) to resolve allegations of false labeling; (vii) to prevent fraud and theft of food items; (viii) to improve food distribution systems and reduce food wastage; and (ix) to help enforce and support meat hygiene in the processing and handing (Furness and Osman, 2006). Although the first is the most critical in minimizing an event from impacting both the public and the industry, the others are also important to justify a comprehensive and effective traceability system.

Traceability systems are required to provide a clear, uninterrupted means of physically tracking an item and/or its constituent components through interlinking nodes of a supply chain, i.e., points in the chain in which the item is processed or handled in some way (Verdenius, 2006). The three main types of information that are collected in traceability systems are location of the product (product flows for individual items or product lots), the condition (how the product was processed and kept during transportation and storage, e.g., modified atmosphere packaging with a certain gas mixture, microbiological limits), and the quality (freshness/taste, storage time limits). Key nodes where information is collected include receipt of raw materials as ingredients of a food, different processing stages ranging from one simple step to complex operations done over a period of time and in different locations, packaging of individual items, boxing and palletizing units, shipping ranging from local delivery to multiple transportation and storage steps for export, storage at retail, and display for the consumer.

The Mechanisms

All traceability systems should exhibit common structural features (Furness and Osman, 2006):

1. item identification for accommodating processing and handling in the supply chain

2. item-attendant and/or item-associated information appropriate to nodal transforms and transactions and any internodal events that have a bearing on traceability
3. process-based information relating to items processed or handled in the supply chain
4. communication links to allow access and exchange of information

Communication information can include data on the contents of a shipment along with the dispatch time, the destination, the intended delivery time, and specific consignment storage and handling details. One essential element in traceability schemes is the need for sufficient information at each node at least one step back and one step forward in the supply chain, so that there is a continuity in the tracking system. Any break in this in a complex food chain means that there may be difficulty finding the source and extent of a problem, whether it is a delivery issue or a contamination one. Identification of items can be at the molecular level such as DNA, protein or metabolic fingerprinting (genomics, proteomics, and metabolomics), or some kind of chemical or biological profile that is generally recognized, e.g., meat animal type, fruit species and variety. Associated with this identification is information on the origin of the specific food item, e.g., field or row location, GM crop, organically certified products.

Code information has to be machine readable and consistent with existing standards for codes, usually alphanumeric (Furness and Osman, 2006). An example of such a code is the Global Trade Item Number (GTIN), which has 14 digits from which a family of four unique numbering structures is derived. These include digits for country, company, item reference, and an error check digit. There are other codes that are regularly used, such as the 18-digit Serial Shipping Container Code for transport data, and a Global Location Number Code that contains unique information on a company and warehouses with contact and bank account information, delivery requirements or restrictions. Data carrier technologies have been developed to give flexibility to traceability systems. These include linear and two-dimensional bar codes, contact and non-contact magnetic data carriers, and radiofrequency identification data (RFID) carriers. Coupled with these are locating mechanisms such as real-time locating systems (RTLS) and global-positioning systems (GPS), and wireless local area network technologies for communications, as well as sensory (quality/spoilage monitoring) and security (fraud and breach in security detection) technologies. An issue with all codes is that they have to be accessible to be readable, especially bar codes, which require optical scanners. Errors or damage to linear bar codes can be overcome with a human operator keying in information. This, however, is more difficult with the more complex coding technologies.

Currently, RFID uses a wireless system that helps with tracking of products, parts, expensive items, and temperature- and time-sensitive goods. RFID chips are small and relatively low-cost circuits capable of communicating with a fixed or portable device, the reader. In order to do so, an antenna, usually made of tin foil, must be attached to the silicone chip. Chip and antenna together are referred to as RFID tags or transponders. RFID tags can be attached to consumer goods, packaging, and other items, and they can also be implanted into animals and even humans (EurActiv Network, 2007). Passive RFID tags do not need a power supply of their own; the minute tension induced from a radiofrequency signal emitted by the reader is sufficient to activate their circuit and to send out short digital information streams in response. Typically, this information consists of a unique identification number that points to an entry in a data base. Semipassive RFID tags have built-in batteries and do not require energy induced from the reader to power the microchip. This allows them to function with much lower signal power levels and over greater distances than passive tags. They are, however, considerably more expensive. Active RFID tags have an on-board power supply, usually a battery, of their own. This allows for more complex circuits to be powered and for more functionality.

Each RFID tag will identify itself when it detects a signal from a reader that emits a radio frequency transmission. When these tags pass through a field generated by a compatible reader, they transmit this information back to the reader, thereby identifying the object. Unlike bar codes, RFID chips can be accessed through normal packaging materials. An advance for a new generation of tags is the ability to read a number of them at once without overwhelming the reader operators. Also, in the near future, processors will attach the tag inside packages of food, and the sensor on the RFID tag will automatically take temperature, moisture, and light incidence readings during intervals, and record this information on the tag (Reynolds, 2007). At any point during the supply chain, the condition of the packaged food and its storage conditions will then be available to processors.

A disadvantage to the use of RFIDs is that they are more expensive than bar codes and so, to date, are only considered for the higher cost items and bulk containers. Also, the error read rates and privacy concerns have held back the development of the technology. Another issue is that RFID does not yet have a standard compared with bar codes that do. Active tags can only operate in the UHF and microwave spectrums; however, there are five frequency bands currently used: low frequency (LF), high frequency (HF), ultra-high frequency (UHF), and microwave or super-high frequency (SHF) bands, but not all of these are usable worldwide (EurActiv Network, 2007). The higher the frequency, the longer the reading distance is, from a few centimeters to several hundred meters. For supply chain purposes, the two frequency bands (860 to 960 MHz and 433 MHz) in the UHF spectrum are most important.

A global product coding standard is being developed by EPCglobal and will be royalty free. This will combine RFID technology, existing communications network infrastructure, and a system called Electronic Product Code (EPC), a number for uniquely identifying an item. EPC was initiated in 2003 with a 96-bit code (Furness, 2006; Vernède and Wienk, 2006) and is being tested for its effectiveness. The development of common communication standards on product and shipment data will allow companies and regulators to share such information, thus speeding up the supply chain and cutting down on errors. The first phase of the testing project will aim to demonstrate the interoperability of the code among multiple trading partners and service providers in a global supply chain. A unified data system would allow changes in information about product sizes, weight, name, price, classification, transport requirements, and volumes to be immediately transmitted along the supply chain.

In Europe, the implementation of the general food law (EU, 2002) and in the United States, the 2002 Bioterrorism Act (FDA, 2002) both require traceability systems for trade purposes and involve all those in the food production system (farm to fork). The Bioterrorism Act requires domestic and foreign facilities that manufacture/process, pack, or hold food for human or animal consumption in the United States to register with the FDA. This helps the FDA determine the location and source of the event and permits the agency to notify quickly those facilities that may be affected in the event of a potential or actual bioterrorism incident or an outbreak of food-borne illness. It is likely that traceability will become increasingly important for trade between countries including the United States through existing and future trade agreements such as WTO, the Common Market for the Southern Cone (MERCOSUL), the North American Free Trade Agreement (NAFTA), and the Free Trade Area of the Americas (FTAA). For animals, livestock producers will likely need to identify each animal through to the end of the production chain by attaching RFID chips to the animal either under the skin, to an ear, or in the ruminant forestomach. For fruits and vegetables, however, it is too expensive or difficult to tag every individual item, and instead information flow from the exporting country to the importing country will occur through tags attached to pallets. Nevertheless, even if the product has complied with all the GAPs, HACCP plans, and required standards of the firms and exporting and the importing countries, the exported product may not be free of pathogens, as illustrated in Chapter 3 with the discussion of outbreaks associated with imported food items. Thus, the traceability systems will have to become even more sophisticated to monitor for potential problems at the different stages from the production on the farm to the retail store in order to identify and flag situations that could lead to microbial contamination or growth.

Another issue is what to do with the multitude of results being transmitted, especially as the tags get down to retail store levels. Action must follow any out-of-compliance situation. Does this include checking on a loss of cold chain during a transportation shipment at sea or in the air to try to save the product, or waiting until it arrives and discarding it for not meeting the importers' criteria? Will governments install their own chips to monitor products for conforming to standards and other criteria and select those lots that have a history of compliance issues for further inspection?

Traceability has always had the attention of discriminating food importers, but it seems likely that before long it will be mandatory for all food producers, processors, and retailers to have minimum standards in tracking ingredients and products and that these systems will be compatible. Also, they will become sufficiently sophisticated to monitor many environmental conditions that impinge on quality and safety of products, and give alerts when there are potential failures. There will be integration between tracking devices and packaging and transport equipment to allow the best design for particular scenarios in farm-to-fork supply systems. They will help reduce human error and limit the ability to tamper with products, and make smuggling and fraud more difficult.

CONCLUDING COMMENTS

For centuries, new and interesting foods have been one of the drivers for exploration and trade, and this trend continues at an ever-increasing pace today. Goods from all over the country and abroad go through regional distribution centers, many of them owned by the large national or international chains. A typical center handles over half a million cases of food items weekly, loading upward of a thousand trucks. A modern superstore has from 30,000 to 50,000 product lines. Since many of these products are imported, the volume of international trade today is staggering. Retailers try to ensure that the foreign supplies are safe and wholesome, but regular recalls of foods, mainly from developing countries, and adverse publicity, such as the recent Chinese contaminated pet food event, put the careful purchaser on guard to be checking on the suppliers of their food and their ingredients. However, the economics of the food supply where there is little markup, dictates that the cheapest sources of food ingredients are going to be sought after even if the risks of contamination may be greater. Thus, if company representatives cannot be present, a third-party certifier in the exporting country may be hired to ensure that the shipment meets both national and company standards, and that good agricultural and manufacturing practices are followed.

The Codex Alimentarius Commission attempts to standardize approaches to the safety of food throughout the world by having all member countries agree on general principles and guidelines for specific commodities and their production. Typically, at ports of entry, companies and countries with a poor record of rejections will have their products inspected more often than others. However, relatively few shipments are physically examined, and considerably fewer samples are taken to be microbiologically tested. Review of the accompanying paper work for correct completion of forms and the appropriateness of the attached HACCP plans are all that may be possible. In addition, smuggling of high-value products because of local demand may obviate the normal routes and present risks to consumers such as the spread of H5N1 virus on poultry carcasses. Countries are increasingly adopting a risk-based approach to monitoring the imported food supply, determining the risk of contamination and illness from the products in question and putting in place a risk ranking to prioritize the most effective use of customs and inspection resources. In some cases the risks will be deemed lower than previously thought. In other cases, the product will require more intensive testing than before. Countries have used safety issues as trade barriers in the past, and the SPS Agreement is supposed to force countries to adopt a risk assessment approach for risks to human health before rejecting a shipment. However, countries have different standards for products such as poultry, and some periodically turn back shipments when the enforcement of these standards is applied even though pathogens such as *Salmonella* and *Campylobacter* cannot be completely eliminated under today's production system.

Another recent issue is the risk of deliberate attack using the food supply. The FDA's Bioterrorism Act is supposed to carefully alert customs officials of shipments before they arrive and have specific paperwork ready for inspection. Included in this is the requirement for exporters to document their suppliers and where the off-loaded products are going. More extensive traceability is being considered by the larger retail operations to quickly locate a problem and contain it; this will be made easier as electronic chips like RFID are fine-tuned and become less expensive. Such tracking systems are good also for reducing food loss through spoilage, for conducting targeted recalls if required, and also to reduce the likelihood of fraud. For example, on 1 September 2007, China started using a new labeling system for food exports that have passed inspections to help trace potential quality problems on record and prevent fake exports, as had occurred. Another reason for good inventory records is the increasing public demand for the country of origin (COOL) on products they purchase, and whether they are organic, wild, or locally produced. Programs to improve the safety of imported food are a mixture of risk-based approaches such as reducing the likelihood of pathogens being present and meeting consumer demands for quality products.

REFERENCES

Alberts, B. 2005. Modeling attacks on the food supply. *Proc. Natl. Acad. Sci. USA* **102:** 9737–9738.

Anonymous. 1991. Cholera—New Jersey and Florida. *Morb. Mort. Wkly. Rep.* **40**(17):287–289.

Anonymous. 1993. U.S. cleared in '89 scare over Chilean fruit. *New York Times.* January 2. [Online.] http://query.nytimes.com/gst/fullpage.html?sec=health&res= 9F0CEEDB1439F931A35752C0A965958260&partner=rssnyt&emc=rss. Accessed 1 June 2007.

Anonymous. 2006. Avian influenza, poultry vs. migratory birds (32) Fri, 14 Jul 2006 23:22:28–0400 Pro Med. [Online.] http://list.uvm.edu/cgi-bin/wa?A2=ind0607B&L= safety&P=6921. Accessed 8 June 2007.

Anonymous. 2007. Te Mata cheese thirsty for raw milk. *Hawke's Bay Today.* May 17. [Online.] http://www.hbtoday.co.nz/localnews/storydisplay.cfm?storyid=3734274&thesection= localnews&thesubsection=&thesecondsubsection=. Accessed 18 May 2007.

Aust, S. D., C. D. Millis, and L. Holcomb. 1987. Relationship of basic research in toxicology to environmental standard setting: the case of polybrominated biphenyls in Michigan. *Arch. Toxicol.* **60:**229–237.

Australian Quarantine and Inspection Service (AQIS). 2002. Accessing the Australian market—Australia's imported food requirements. [Online.] http://www.daff.gov.au/aqis/import/ food/notices. Accessed 1 April 2008.

Beato, M. S., C. Terregino, G. Cattoli, and I. Capua. 2006. Isolation and characterization of an H10N7 avian influenza virus from poultry carcasses smuggled from China into Italy. *Avian Pathol.* **35:**400–403.

Board on Global Health. 2006. Addressing Foodborne Threats to Health: Policies, Practices, and Global Coordination. Forum on Microbial Threats. Workshop Summary. Institute of Medicine of the National Academies, The National Academies Press, Washington, DC.

Bowers, J., A. Dalsgaard, A. DePaola, I. Karunasagar, T. McMeekin, M. Nishibuchi, K. Osaka, J. Sumner, and M. Walderhaug. 2005. FAO/WHO Microbiological Risk Assessment Series, no.9. pp. 1–109. Risk assessment of choleragenic *Vibrio cholerae* O1 and O139 in warmwater shrimp in interational trade. World Health Organization, Geneva, Switzerland.

Buchanan, R., R. Lindqvist, T. Ross, M. Smith, E. Todd, and R. Whiting. 2004. Risk assessment of *Listeria monocytogenes* in ready-to-eat foods, pp. 1–307. Technical Report. FAO/WHO Microbiological Risk Assessment Series, no.5. World Health Organization, Geneva, Switzerland.

Bull A. L., S. K., Crerar, and M. Y. Beers. 2002. Australia's Imported Food Program—a valuable source of information on micro-organisms in foods. *Commun. Dis. Intell.* **26**(1). [Online.] http://health.gov.au/internet/wcms/publishing.nsf/Content/cda-cdi2601-pdf-cnt.htm/ $FILE/cdi2601f.pdf. Accessed 11 November 2007.

Butler, R. J., J. G. Murray and S. Tidswell. 2003. Quality assurance and meat inspection in *Aust. Rev. Sci. Tech. Off. Int. Epiz.* **22:**697–712.

Canadian Food Inspection Agency (CFIA). 2007. Melamine in imported products. [Online.] http://www.inspection.gc.ca/english/fssa/concen/specif/vegproe.shtml. Accessed 11 June 2007.

Caswell, J. A., and S. Joseph. 2008. Consumer demand for quality: major determinant for agricultural and food trade in the future? *J. Int. Agric. Trade Dev.* 4(1): 99–116.

Center for Infectious Disease & Research Policy (CIDRAP), University of Minnesota News. 2006. Michigan officials track smuggled Chinese poultry. {online.] http://www.cidrap.umn.edu/cidrap/content/influenza/avianflu/news/jul1306avian.html. Accessed 21 May 2007.

Chung, A. 2007. The trouble on the world's farm. 20 May. *Toronto Star* [Online.] http://www.thestar.com/News/article/215857. Accessed 11 June 2007.

Codex Alimentarius Commission (CAC). 1979. Recommended International Code of Hygienic Practice for Foods for Infants and Children. CAC/RCP 21-1979.

Codex Alimentarius Commission (CAC). 1997a. Guidelines for the Design, Operation, Assessment and Accreditation of Food Import and Export Inspection and Certification Systems. CAC/GL 26-1997. [Online.] www.codexalimentarius.net/download/standards/362/CXG_034e.pdf. Accessed 11 November 2007.

Codex Alimentarius Commission (CAC). 1997b. Codex Principles for the Establishment and Application of Microbiological Criteria for Foods. CAC/GL 21-1997. [Online.] www.codexalimentarius.net/download/standards/394/CXG_021e.pdf. Accessed 11 November 2007.

Codex Alimentarius Commission (CAC). 2002. Report of the Evaluation of the Codex Alimentarius and Other FAO and WHO Food Standards Work. [Online.] http://www.fao.org/docrep/meeting/005/y7871e/y7871e00.htm. Accessed 11 November 2007.

Codex Alimentarius Commission (CAC). 2004. General Guidelines on Sampling. CAC/GL 50-2004. [Online.] www.codexalimentarius.net/download/standards/10141/CXG_050e.pdf. Accessed 11 November 2007.

Codex Alimentarius Commission (CAC). 2006. Codex Standard for Whey Cheeses (Codex Stan A-7-1971, Rev. w-2006). [Online.] http://www.codexalimentarius.net/web/standard_list.do?lang=en. Accessed 11 November 2007.

Cody, E. 2007. China contests complaints on exported food. *Washington Post* [Online.] http://www.washingtonpost.com/wp-dyn/content/article/2007/05/31/AR2007053100901.html. Accessed 31 May 2007.

Dagg, P. J., R. J. Butler, J. G. Murray, and R. R. Biddle. 2006. Meeting the requirements of importing countries: practice and policy for on-farm approaches to food safety. *Rev. Sci. Tech.* 25:685–700.

Ehmke, M. T. 2006. International differences in consumer preferences for food country-of-origin: a meta-analysis. Selected paper presented at the annual meeting of the American Agricultural Economics Association, Long Beach, CA, 23 to 26 July 2006. http://ageconsearch.umn.edu/bitstream/123456789/14638/1/sp06eh01.pdf. Accessed 2 April 2008.

ElAmin, A. 2007. Bernard Matthews lays off workers due to bird flu. February 21. MeatProcess.com. [Online.] http://www.meatprocess.com/news/ng.asp?id=74385-bernard-matthews-bird-flu-h-n. Accessed 26 June 2007.

EurActiv Network. 2007. Radio Frequency Identification Chips (RFID). [Online.] http://www.euractiv.com/en/infosociety/radio-frequency-identification-chips-rfid/article-158701. Accessed 22 August 2007.

European Food Safety Authority (EFSA) Task Force. 2007. Report of the Task Force on Zoonoses Data Collection on the analysis of the baseline survey on the prevalence of *Salmonella* in broiler flocks of *Gallus gallus*, in the EU, 2005–2006 [1]—Part A: *Salmonella* prevalence estimates. *EFSA J.* **98**:1–85. [Online.] http://www.efsa.europa.eu/en/science/ monitoring_zoonoses/reports/zoon_report_finbroilers.html. Accessed 22 June 2007.

European Union (EU). 2002. Regulation (EC) no. 178/2002 laying down the general principles and requirements of food law, establishing the European Food Safety Authority and laying down procedures in matters of food safety. *Official Journal of the European Union* no. L 031, 1.02.2002, p. 1.

Eyre, C. 2007. Australia gives go-ahead for GM wheat testing. FoodNavigatorU.S.A.com. [Online.] http://www.foodnavigator-usa.com/news-by-product/news.asp?id=77490&idCat= 0&k=Victorian-Department-of-Primary-Industri—GM-wheat Accessed 22 June 2007.

Food & Water Watch. 2007. Import alert: government fails consumers, falls short on seafood inspections. [Online.] http://www.foodandwaterwatch.org/press/publications/reports/ import-alert. Accessed 31 May 2007.

Food Standards Australia New Zealand (FSANZ). 2003. First Review Report Proposal P239. 19 March. 08/03 *Listeria* risk assessment and risk management strategy. [Online.] http://www.foodstandards.gov.au/_srcfiles/P239_Listeria_FRR.doc. Accessed 18 May 2007.

Furness, A. 2006. Data carriers for traceability. *In* I. Smith and A. Furness (ed.), *Improving Traceability in Food Processing and Distribution*, p. 199–237. CRC Press, Woodhead Publishing Ltd, Cambridge, England.

Furness, A., and K. A. Osman. 2006. Developing traceability systems across the food supply chain: an overview. p. 3–25. *In* I. Smith and A. Furness (ed.), *Improving Traceability in Food Processing and Distribution*. CRC Press, Woodhead Publishing Ltd, Cambridge, England.

Garcia, M., A. Fearne, J. A. Caswell, and S. Henson. 2007. Co-regulation as a possible model for food safety governance: opportunities for public-private partnerships. *Food Policy* **32**: 299–314.

General Accounting Office (GAO). 1998. Federal Efforts to Ensure Imported Food Safety Are Inconsistent and Unreliable. Statement of Robert E. Robertson, Associate Director, Food and Agriculture Issues, Resources, Community, and Economic Development Division. GAO/ T-RCED-98-191. U.S. General Accounting Office, Washington, DC.

General Accounting Office (GAO). 2003. Country-of-Origin Labeling: Opportunities for U.S.DA and Industry to Implement Challenging Aspects of the New Law. U.S. General Accounting Office, Washington, DC.

General Accountability Office (GAO). 2007. Federal Oversight of Food Safety: High-Risk Designation Can Bring Attention to Limitations in the Government's Food Recall Programs. GAO-07-785T. 24 April 2007. U.S. General Accounting Office, Washington, DC.

Golan, E., T. Roberts, E. Salay, J. Caswell, M. Ollinger, and D. Moore. 2004. *Food Safety Innovation in the United States: Evidence from the Meat Industry*. Agricultural Economic Report no. 831. Department of Agriculture, Economic Research Service, Washington, DC. [Online.] http://www.ers.usda.gov/publications/aer831/. Accessed 11 November 2007

Gorham, B. 2007. U.S. group claims banned Canadian cows moving across the border. The Globe and Mail, *Canadian Press* June 15. [Online.] http://www.theglobeandmail.com/ servlet/story/RTGAM.20070615.wcows0615/BNStory/National/. Accessed 22 June 2007.

Hald, T., D. Vose, H. C. Wegener, and T. Koupeev. 2004. A Bayesian approach to quantify the contribution of animal-food sources to human salmonellosis. *Risk Anal.* 24:255–269.

Hatanaka, M., C. Bain, and L. Busch. 2005. Third-party certification in the global agrifood system. *Food Policy* 30:354–369.

Hedges, S. 2007. Mad cow 'minimum risk' from Canadian beef, U.S. says. *Chicago Tribune* 15 January 2007. [Online.] http://www.chicagotribune.com/news/custom/newsroom/chi-070115madcow,1,4468187.story. Accessed 22 June 2007.

Heller, L. 2007. Global trade deal looks cloudy as Doha collapses—again. FoodNavigatorEurope.com. [Online.] http://www.foodnavigator.com/news/ng.asp?n=77587&m=1fne622&c=epfadfzbkiaduyg. Accessed 22 June 2007.

Hennesy, T. W., C. W. Hedberg, L. Slustker, K. E. White, J. M. Besser-Wiek, M. E. Moen, J. Feldman, W. W. Coleman, L. M. Edmonson, K. L. MacDonald, and M. T. Osterholm. 1996. A national outbreak of *Salmonella enteritidis* infections from ice cream. *N. Engl. J. Med.* 334:1281–1286.

Holtby, I., G. M. Tebbutt, S. Anwar, J. Aislabie, V. Bell, W. Flowers, J. Hedgley, and P. Kelly. 2006. Two separate outbreaks of *Salmonella* enteritidis phage type 14b food poisoning linked to the consumption of the same type of frozen food. *Public Health* 120:817–823.

Huss, H. H., L. Ababouch, and L. Gram. 2004. Assessment and management of seafood safety and quality. FAO Fisheries Technical Paper 444. Food and Agriculture Organization, Rome, Italy.

International Association for Food Protection (IAFP). 2007. *Procedures to Investigate Foodborne Illness*, 5th ed. 1999, revised 2007. International Association for Food Protection, Des Moines, IA.

Jaffee, S., and S. Henson. 2004. Standards and Agro-Food Exports from Developing Countries: Rebalancing the Debate. Policy Research Working Paper 3348. The World Bank, Washington, DC.

Josling, T., D. Roberts, and D. Orden. 2004. *Food Regulation and Trade: Toward a Safe and Open Global System*. Institute for International Economics, Washington, DC.

Kinde, H., A. Mikolon, A. Rodriguez-Lainz, C. Adams, R. L. Walker, S. Cernek-Hoskins, S. Treviso, M. Ginsberg, R. Rast, B. Harris, J. B. Payeur, S. Waterman, and A. Ardans. 2007. Recovery of *Salmonella, Listeria monocytogenes,* and *Mycobacterium bovis* from cheese entering the United States through a noncommercial land port of entry. *J. Food Prot.* 70:47–52.

Lusk, J. L., J. Brown, M. Tyler, I. Proseku, R. Thompson, and J. Welsh. 2006. Consumer Behavior, Public Policy, and Country-of-Origin Labeling. *Rev. Agric. Econ.* 28:284–292.

Matthews, K. H., J. Bernstein, and J. C. Buzby. 2003. International trade of meat/poultry products and food safety issues. p. 48–73. *In* J.C. Buzby (ed.), *International Trade and Food Safety: Economic Theory and Case Studies*. U.S. Department of Agriculture, Economic Research Service, Washington DC.

Mercer, C. 2005. Australia lifts Roquefort cheese safety ban. [Online.] http://www.ap-foodtechnology.com/news/ng.asp?id=62799-fsanz-roquefort-speciality-cheese. Accessed 18 May 2007.

National Academies, Institute of Medicine. 2007. *Seafood Choices: Balancing Benefits and Risks*. The National Academies Press, Washington, DC.

National Marine Fisheries Service (NMFS). 2007. FishWatch: US Seafood Facts, Trade. [Online.] http://www.nmfs.noaa.gov/fishwatch/trade_and_aquaculture.htm. Accessed 25 September 2007.

Oliver, J. D., and J. B. Kaper. 2007. *Vibrio* species. p. 343–379, *In* M. P. Doyle and L. R. Beuchat (ed.), *Food Microbiology: Fundamentals and Frontiers*, 3rd ed. American Society for Microbiology Press, Washington, DC.

Reynolds, G. 2006. FSA stands by Spanish Salmonella claims. FoodQualitynews.com. November 20. [Online.] http://www.foodqualitynews.com/news/ng.asp?n=72169-fsa-salmonella-eggs. Accessed 22 August 2007.

Reynolds, G. 2007. German project to create meat analysis RFID. *FoodProduction Daily* [Online.] http://www.foodproductiondaily.com/news/ng.asp?id=73150-rfid-izm-meat. Accessed 8 June 2007.

Ross, T., P. Dalgaard, and S. Tienungoon. 2000. Predictive modelling of the growth and survival of *Listeria* in fishery products. *Int. J. Food Microbiol.* **62**:231–245.

Sinclair, S. 2007. NAFTA Chapter 11 Investor-State Disputes (to 1 March 2007). Trade and Investment Research Project, Canadian Centre for Policy Alternatives. [Online.] http://policyalternatives.ca/documents/National_Office_Pubs/2007/NAFTA_Dispute_Table_March2007.pdf. Accessed 26 June 2007.

Smith DeWaal, C., K. Johnson, and F. Bhuiya. 2006. *Outbreak Alert: Closing the Gaps in Our Federal Food-Safety Net.* Center for Science in the Public Interest, Washington, DC.

Sterns, P. A., J-M. Codron, and T. Reardon. 2001. Quality and quality assurance in the fresh produce sector: a case study of European retailers. Paper presented at American Agricultural Economics Association annual meeting, Chicago IL, 5–8 August. [Online.] http://agecon.lib.umn.edu/cgi-bin/pdf_view.pl?paperid=2770&ftype=.pdf. Accessed 11 November 2007

Souza-Monteiro, D., and J. A. Caswell. 2006. Traceability adoption at the farm level: analysis of the Portuguese pear industry. Paper presented at American Agricultural Economics Association annual meeting, Long Beach, Calif., July. [Online.] http://agecon.lib.umn.edu/cgi-bin/pdf_view.pl?paperid=21673&ftype=.pdf. Accessed 11 November 2007.

Tan, C. G., H. S. Sandhu, D. C. Crawford, S. C. Redd, M. J. Beach, J. W. Buehler, E. A. Bresnitz, R. W. Pinner, B. P. Bell, and the Regional Anthrax Surveillance Team. 2002. Surveillance for anthrax cases associated with contaminated letters, New Jersey, Delaware, and Pennsylvania, 2001. *Emerg. Infect. Dis.* **8**(10). [Online.] http://www.cdc.gov/ncidod/EID/vol8no10/02-0322.htm. Accessed 25 June 2007.

Todd, E. C. D. 2007. *Listeria*: risk assessment, regulatory control and economic impact, p. 717–762. *In* E. T. Ryser and E. Marth (ed.), Listeria, *Listeriosis and Food Safety*, 3rd ed. Taylor and Francis, Boca Raton, Fla.

Todd, E. C. D, C. K. Harris, A. J. Knight, and M. R. Worosz. 2007. Spinach and the media: how we learn about a major outbreak. *Food Prot. Trends* **27**:314–321.

Unterschultz, J., K. Quagrainie, and M. Veeman. 1998. Effects of product origin and selected demographics on consumer choice of red meats. *Can. J. Agric. Econ.* **46**:201–219.

U.K. Department for Environment, Food, Rural Affairs (DEFRA). 2007. Summary of initial epidemiological and virological investigations to determine the source and means of introduction of highly pathogenic H5N1 avian influenza virus into a turkey finishing unit in Suffolk, as at 14 February 2007. [Online.] http://www.defra.gov.uk/animalh/diseases/notifiable/disease/ai/pdf/epidemiological160207.pdf. Accessed 26 June 2007.

U.K. Food Standards Agency (FSA). 2004. Action stepped up on *Salmonella* outbreaks. Food Standards Agency. [Online.] http://www.food.gov.uk/news/newsarchive/2004/oct/spanisheggs. Accessed 22 June 2007.

U.K. Food Standards Agency (FSA). 2006. FSA surveys non-UK eggs for *salmonella*, Wednesday 15 November 2006. [Online.] http://www.food.gov.uk/news/newsarchive/2006/nov/eggs. Accessed 22 August 2007.

U.S. Food and Drug Administration (FDA). 2002. Public Health Security and Bioterrorism Preparedness and Response Act of 2002. [Online.] http://www.fda.gov/oc/bioterrorism/bioact.html. Accessed 26 June 2007.

U.S. Food and Drug Administration (FDA). 2003. Risk Assessment for food terrorism and other food safety concerns. Center for Food Safety and Applied Nutrition/Office of Regulations and Policy, Food and Drug Administration. [Online.] http://www.cfsan.fda.gov/~dms/rabtact.html#ftnref46. Accessed 26 June 2007.

U.S. Food and Drug Administration (FDA). 2007a. CARVER + Shock Software Tool, Food Defense and Terrorism, Food and Drug Administration. Center for Food Safety and Applied Nutrition, Food and Drug Administration. [Online.] http://www.cfsan.fda.gov/~dms/carver.html. Accessed 25 June 2007.

U.S. Food and Drug Administration (FDA). 2007b. FDA detains imports of farm-raised Chinese seafood. *FDA News* 28 June 2007. [Online.] http://www.fda.gov/bbs/topics/NEWS/2007/NEW01660.html. Accessed 11 November 2

U.S. Food and Drug Administration (FDA). 2008. Compliance policy guide. Sec. 555.320. *Listeria monocytogenes.* http://www.fda.gov/ora/compliance_ref/cpg/cpgfod/draft_cpg555-320.html. Accessed 2 April 2008.

van Larebeke, N., L. Hens, P. Schepens, A. Covaci, J. Baeyens, K. Everaert, J. L. Bernheim, R. Vlietinck, and G. De Poorter. 2001. The Belgian PCB and Dioxin incident of January–June 1999: Exposure data and potential impact on health. *Environ. Health Perspect.* **109:**265–273.

Verdenius, F. 2006. Using traceability systems to optimize business performance, p. 26–51. *In* I. Smith and A. Furness (ed.), *Improving Traceability in Food Processing and Distribution.* CRC Press, Woodhead Publishing Ltd, Cambridge, England.

Vernède, R., and I. Wienk. 2006. Storing and transmitting traceability data across the food supply chain, p. 183–198. *In* I. Smith and A. Furness (eds.), *Improving Traceability in Food Processing and Distribution.* CRC Press, Woodhead Publishing Ltd, Cambridge, England.

Wein, L. M., and Y. Liu. 2005. Analyzing a bioterror attack on the food supply: the case of botulinum toxin in milk. *Proc. Natl. Acad. Sci. USA* **102:**9984–9989.

Weiss, R. 2007. *The Washington Post* May 20. [Online.] http://www.washingtonpost.com/wp-dyn/content/article/2007/05/19/AR2007051901273.html. Accessed 21 May 2007.

World Health Organization (WHO). 2002. Terrorist Threats to Food: Guidance for Establishing and Strengthening Prevention and Response Systems. Food Safety Department,

World Health Organization, Geneva, Switzerland. [Online.] http://www.who.int/foodsafety/publications/general/en/terrorist.pdf. Accessed 25 June 2007.

World Bank. 2005. Food Safety and Agricultural Health Standards: Challenges and Opportunities for Developing Country Exports. Report no. 31027, Poverty Reduction & Economic Management Trade Unit, Agriculture and Rural Development Department, The World Bank, Washington, DC.

Summary and Perspective of the Impact of Imported Foods on the Microbiological Safety of the United States' Food Supply

Marilyn C. Erickson and Michael P. Doyle

Globalization of the food supply has for the most part benefited both exporting and importing countries. Many of the exporting countries are developing nations, and their participation in the global marketplace has generated an influx of revenues that boost their economy and per capita income. For the United States, the importation of foods has provided diversity in selection, year-round supply, and low costs. In Chapter 1, foods imported into the United States were reviewed with particular attention to their percentage of total food consumed, types of foods being imported, and the major exporting countries for different commodities. Noted was the continued growth in consumption of imported foods over the past two decades, with both short- and long-term projections suggesting that this trend will continue.

Growth of foods imported into the United States is not equitably distributed across all commodities. Instead, growth has occurred largely in the seafood, red meat, vegetables, and fruits/tree nuts commodity areas, with current volumes accounting for 80%, 10%, 9%, and 36% of total domestic consumption, respectively. Within these food groups, importation of specific food products has vacillated in their share of domestic consumption. For example, the average share of imported pork has declined since 1980 while the share of beef, lamb, and mutton has increased. For vegetables, there has been a rapid increase in importation of fresh vegetables, whereas processed vegetables have grown only slightly.

A variety of supply- and demand-side drivers are responsible for the shifts in food items being imported into the United States. Supply-side drivers that occur in other countries are low land values, water availability, and appropriate

MARILYN C. ERICKSON AND MICHAEL P. DOYLE, Center for Food Safety and Department of Food Science and Technology, University of Georgia, 1109 Experiment Street, Griffin, GA 30223-1797.

climate conditions for the food's production. In addition, labor costs are one of the chief supply-side drivers for the relocation of fisheries and fruit and vegetable production to other countries. There are vast differences in labor costs between the United States and many other countries. For example, wages in Brazil and Mexico were 17 and 11%, respectively, of those in the United States (U.S. Department of Labor, 2006). In rural areas of China, wage rates are as low as $2 per day (Huang and Gale, 2006). Moreover, in the United States, documented and undocumented immigrants account for a large percentage of the agricultural workforce. The government's efforts to cut down on illegal immigration could accelerate the relocation of agricultural production to other countries.

Another supply-side driver that has influenced and will continue to influence the surge of food exports to the United States is export-oriented growth policies adopted by less developed countries. As a result of these policies, both China and Brazil have become major food exporters to the United States. For countries whose economies depend heavily on agriculture, policies supporting food production for overseas markets are unlikely to change in the foreseeable future. Furthermore, changes to governmental policies within the United States could impact tremendously the types of food products being imported. For example, there has been discussion for many years to eliminate farm subsidies in the Farm Bill because its original purpose is no longer valid. If the Farm Bill is discontinued at any point in the future, elimination of price supports may lead to the importation of commodities not currently being imported.

Demand-side drivers are also critical factors influencing the increased demand for imported food. In particular, changes in the consumption patterns of Americans have created a demand for convenience foods and a year-round supply of seasonally based fruits and vegetables. Increased travel experiences and marketing on the Internet have also been conducive to familiarizing the consumer with new food products that are available only through importation. Moreover, ethnic communities create niche markets for many of the imported seafood, nut, fruit, and vegetable items.

Countries that are the major exporters to the United States vary with the specific commodity. Commodities that are more expensive, such as salmon, tend to be from countries that have a relatively high per capita income, whereas commodities such as tilapia, which is a less expensive item, originate in countries with lower per capita incomes. With this type of disparity, Canada has had a greater impact on the United States economically, but China has had a greater proportion of the import volume, which translates to greater product exposure and potential adverse public health risks.

Eight governmental agencies have a role in monitoring the large volumes of food imported into the United States, but the Food and Drug Administration

(FDA) has the primary responsibility, with oversight of about 80% of the food supply. The U.S. Department of Agriculture Food Safety Inspection Service (USDA-FSIS) has the second largest role, being responsible for imported meat, poultry, and some egg products. Our current food import regulatory system has two major weaknesses: (i) there are a number of redundancies in the import procedures because of overlapping requirements for specific food products, and (ii) different statutory authority exists for USDA-FSIS-regulated food products and FDA-regulated food products, with USDA-FSIS imposing an equivalency requirement for countries exporting foods to the United States and FDA not having this statutory authority but instead relying on inspections at the ports of entry. Hence, the FDA has emphasized product testing as its approach to verifying the safety of imported food products, whereas USDA-FSIS has emphasized process oversight. FDA's method for regulating imported foods has been a major point of concern given this agency's limited resources. There are 326 ports in the United States through which FDA-regulated products enter the country; however, there are only 380 FDA field personnel that regularly perform inspections at these ports (FDA, 2007a). Hence, only about one percent of the food imported under its jurisdiction is visually checked upon entry into the United States and a much smaller percentage is physically inspected, including some laboratory tests. At face value, this low percentage seems appalling; however, FDA's inspections are risk based, targeting riskier food groups and companies with a history of violations for inspections. The effectiveness of this approach to protecting public health in the face of rising import volume is highly questionable.

Support for the effectiveness of FDA and USDA-FSIS programs for monitoring the safety of imported foods is suggested from the relatively low number of reported outbreaks/illnesses attributed to this segment of foods. However, the number of outbreaks is likely underestimated due to the difficulties of epidemiologic investigations in firmly associating an outbreak with an imported food. In particular, the actual origin of a food may be obscured as domestic and imported products may be mixed together and relabeled by intermediary distributors. Even when an imported food is identified as the vehicle of food-borne illnesses, in many cases, investigators fail to identify the country of origin. When imported foods have been documented as vehicles of outbreaks in the United States, produce and seafood (items regulated by FDA) are the dominant sources, whereas meat (items regulated by USDA-FSIS) has not been identified as a vehicle. This may be due to the large percentage of imported seafood and fresh fruits and vegetables consumed in the United States compared with the relatively low percentage of imported meat and poultry. Another contributing factor could be the differences in regulatory approaches used by USDA-FSIS and FDA to regulate imported foods (i.e., process oversight versus product inspection, respectively).

Several studies have been conducted to identify the microbiological safety of foods in developing countries by assaying different food products for harmful microorganisms and their toxins. Examples include studies of the occurrence of pathogenic bacteria in raw milk and pasteurized milk products in Zimbabwe (Gran et al., 2003), of levels of aflatoxin B_1 in milk in Argentina (Lopez et al., 2003), of the prevalence of *Salmonella* in raw vegetables in Malaysia (Sallel et al., 2003), and of the microbial load of sheep/goat carcasses from an abattoir and traditional meat shops in India (Bhandare et al., 2007). Results revealed high levels of contamination in the Zimbabwe milk products, Malaysian vegetables, and Indian sheep/goat products, indicating that foods produced and processed in those countries could pose a significant health threat to consumers. Similar hazards are likely to also exist in other developing countries that have high mortality rates among children that are attributable to diarrheal diseases (WHO, 2004) and high estimates of occurrences of food-borne infections and intoxications (Henson, 2003). An illustration of the disparities in public health between developing countries and industrialized countries, like the United States, is the comparison of estimated loss of disability-adjusted life years (DALYs) for those countries. For example, in 1990, estimated DALYs for India, China, and Latin American countries were 128, 16, and 23 times greater than DALYs for established market economies of highly industrialized countries (Murray and Lopez, 1996).

Factors contributing to microbiological contamination of foods and associated disease burdens within developing countries include the use of animal and human waste in agricultural enterprises. Within the United States and most other industrialized countries, the use of human wastes or wastewater for edible crops is not allowed or is strictly regulated, and animal wastes are often composted or treated for pathogen inactivation before utilization for production of crops for human consumption. In contrast, practices prevalent within many developing countries include the use of fecal-contaminated irrigation water for fruit and vegetable production, the use of night soil as a soil amendment, and the use of uncomposted chicken manure and human feces in aquaculture production. Such wastes serve as carriers of a variety of pathogens, including enteric bacteria (e.g., *Salmonella*, *Escherichia coli* O157:H7, and *Campylobacter*), viruses, and parasites. Inadequately trained workers, insufficient governmental oversight, lack of basic infrastructures to treat wastewater, and longstanding cultural attitudes of using sustainable agricultural practices that are insanitary are the root causes of these public health threats. These sustainable conditions subsequently create an environment where basic hygienic practices are deficient both by personnel and by processing establishments. Not all developing countries, however, apply such practices that endanger public health. An example is Malaysia, where the basic elements of a national system of food safety control have generally been implemented and efforts focus on upgrading the food safety system to achieve compliance with international standards and/or to develop food safety capacity within the private sector (Henson, 2003).

Unhygienic food production and processing practices used in exporting countries are reflected in the agricultural and food products refused at the U.S. border. These data are routinely published by the FDA on its OASIS website (http://www.fda.gov/ora/oasis/ora_oasis_ref.html). Table 1 summarizes violations associated with microbiological contamination in 2007. During that period of time, total violations were nearly 70% of violations recorded during the period from July 1996 to June 1997 (Unnevehr, 2000). Moreover, more than half of the violations a decade ago were associated with microbiological issues (Unnevehr, 2000), whereas microbiological violations in 2007 ranged from 29 to 38% of the total food import violations. Within the latter time frame, filthy and *Salmonella* violations dominated and represented 54% and 30%, respectively, of the total number of microbiology-related violations. Similarly, filthy and *Salmonella* violations were the major codes cited for seafood refusals; however, the number of violations in 2007 were at 40 and 17% of the levels that occurred in 2001 (Table 2). Although an improved quality in the food being imported could account for the decrease in microbiology-related violations, it is more likely that some microbiologically contaminated products are being rejected for nonmicrobiological violations.

Food groups that had the greatest number of microbiologically associated refusals at the U.S. border in 2007 are listed in Table 3. Given the large volume of imported seafood, it is not surprising that seafood tops the list; however, Indonesia, the sixth leading seafood exporter to the United States (see Table 5, Chapter 1), has had the greatest number of violations associated with seafood-related microbiological issues. In contrast, the major exporter to the United States of vegetables is Mexico and it accounts for 51% of the vegetable-related microbiological refusals.

Spices are another food group with a large number of refusals, which is of particular concern because these products become ingredients for many other food products and hence contamination could be amplified. Although outbreaks of salmonellosis have been associated with seasonings like paprika, using data from the Centers for Disease Control and Prevention National Surveillance System, Vij et al. (2006) did not find any increases in the reported incidence of laboratory-confirmed salmonellosis in states that received spices contaminated with selected rare *Salmonella* serotypes.

Table 1 FDA import refusals by violation code in 2007[a]

Date	Total no. of import violations	No. of microbiological violations (% of total)	Microbial contamination			Microbial by-products		Filthy	Other sanitation/ processing violations[b]
			Salmonella	Listeria	Bacteria	Histamine	Mycotoxin		
Jan 2007	720	236 (32.8)	87	4	0	4	13	95	33
Feb 2007	590	182 (30.8)	57	11	1	0	3	82	28
Mar 2007	707	258 (36.5)	82	1	1	5	10	151	8
Apr 2007	671	239 (35.6)	86	3	3	5	1	121	20
May 2007	1,070	314 (29.3)	96	5	2	14	0	163	34
Jun 2007	465	158 (34.0)	47	0	1	4	2	94	10
Jul 2007	535	190 (35.5)	56	1	3	2	1	116	11
Aug 2007	634	216 (34.1)	66	0	1	1	2	133	13
Sept 2007	545	174 (31.9)	59	4	2	3	0	90	16
Oct 2007	724	276 (38.1)	73	18	4	9	2	154	16
Nov 2007	678	241 (35.5)	48	3	4	1	8	147	30
Dec 2007	722	247 (34.2)	57	13	7	2	2	142	24
12-month totals	8,061	2,731 (33.9)	814	63	29	50	44	1,488	243

[a]Compiled from FDA's OASIS Web site (http://www.fda.gov/ora/oasis/ora_oasis_ref.html) for January to December 2007.
[b]INSANITARY, MFR INSAN, MFRHACCP, IMPTRHACCP, and UNDER PRC.

Table 2 Microbiological violation codes associated with seafood refusals in 2001[a] and 2007[b]

		No. of violations	
Violation code	Violation description	2001	2007
Salmonella	The article appears to contain *Salmonella*, a poisonous and deleterious substance that may render it injurious to health.	1,832	308
Filthy	The article appears to consist in whole or in part of a filthy, putrid, or decomposed substance or be otherwise unfit for food.	1,460	582
Insanitary	The article appears to have been prepared, packed, or held under insanitary conditions whereby it may have been contaminated with filth, or whereby it may have been rendered injurious to health.	351	12
Listeria	The article appears to contain *Listeria*, a poisonous and deleterious substance that may render it injurious to health.	170	22
Histamine	The article appears to contain histamine, a poisonous and deleterious substance that may render it injurious to health.	123	50
IMPTRHACCP	The food appears to have been prepared, packed, or held under insanitary conditions, or may have become injurious to health, due to the failure of the importer to provide verification of compliance.	41	46

[a]Allshouse et al. (2003).
[b]Compiled from FDA's OASIS Web site (http://www.fda.gov/ora/oasis/ora_oasis_ref.html) for January to December 2007.

Table 3 Top five food groups by rank of number of import refusals associated with microbiological issues in 2007[a]

Rank	Food group	Total no. of refusals	Top country (no. of refusals)	No. of countries with refusals
1	Seafood	1,064	Indonesia (202)	51
2	Spices	403	India (235)	24
3	Fruit	301	China (61)	33
4	Vegetables	266	Mexico (136)	28
5	Cheese	203	Mexico (65)	17

[a]Compiled from FDA's OASIS Web site (http://www.fda.gov/ora/oasis/ora_oasis_ref.html) for January to December 2007.

Taking into consideration all food groups when microbiological contaminants serve as the reason for refusal of imported foods, the country with the greatest number of refusals during 2007 was India, followed by Mexico, China, Vietnam, and Indonesia (Table 4). However, Mexico outranks India when all sources of violations (microbiological, chemical, labeling, improper documentation, etc.) are included (Becker, 2007). The dominant reason cited for microbiologically based refusals has depended on the country. For example, India and Vietnam had more *Salmonella* violations than filthy violations, whereas Mexico, China, and Indonesia had more filthy violations than *Salmonella*

Table 4 Top five countries by rank of number of import refusals associated with microbiological issues in 2007[a]

| Rank | Country | Total no. of refusals (% of microbiological violations[b]) | Major violation codes | |
			Filthy (no. of refusals)	*Salmonella* (no. of refusals)
1	India	474 (17.4)	184	278
2	Mexico	398 (14.6)	257	65
3	China	288 (10.5)	258	17
4	Vietnam	218 (8.0)	77	117
5	Indonesia	213 (7.8)	126	64

[a]Compiled from FDA's OASIS Web site (http://www.fda.gov/ora/oasis/ora_oasis_ref.html) for January to December 2007.
[b]The total number of microbiological violations was 2,731.

violations (Table 4). These patterns of association likely reflect the types of major commodities that the country exports to the United States and the specific types of inspection assays performed for different commodities. The large number of microbiologically associated refusals that occur with imported products from developing countries is an issue that warrants attention.

A variety of programs could be applied in developing countries to improve the microbiological safety of the foods they export. Included among these programs are the extensive standards, codes of practice, and guidance on microbiological food safety that the Codex Alimentarius Commission has developed. Criticism of the Codex system has focused on the lack of timeliness and inefficiency of the standard-setting process and on inadequate opportunities for developing countries to participate effectively in the process without additional capacity building. As part of that capacity building, risk management and programs designed to deliver greater and more effective communication are important tools that should be incorporated into production and processing operations of exporting countries. Examples include hazard analysis critical control point (HACCP) systems, sanitary and phytosanitary measures, good agricultural practices, and third-party certifications.

Implementation of risk management programs in developing countries is hampered by two issues: (i) insufficient awareness of the source and severity of food safety and hygiene problems and (ii) lack of basic infrastructure to provide water and sanitation services. To address the first issue, training programs have been implemented by international organizations, such as the Food and Agriculture Organization (FAO) of the United Nations. For example, FAO carried out over 70 workshops and trained more than 1,800 professionals from industry and government in HACCP-based fish quality and safety systems during the past decade (Ababouch, 2006). This training subsequently enabled many developing countries to meet the safety and quality requirements of international markets. Such workshops are helpful but are not the sole solution. Major advances will require a multifaceted approach that addresses the influence of the country's environment, culture, and economics on food safety practices. For example, four years after the conclusion of a $135 million Indonesian Rural Water Supply and Sanitation Sector project, only 30 to 40% of the water and sanitation facilities constructed were still functioning (Asian Development Bank [ADB], 2004). Similarly, <50% of the ≈250,000 hand pumps installed throughout rural Africa are estimated to be operational (Harvey and Reed, 2004); this demonstrates a lack of local incentive to operate and maintain facilities. According to Curtis et al. (2001), change in behavior and adoption of water and sanitation interventions would likely be more effective if built on local research and if locally appropriate channels of communication were used repeatedly and for an extended period of time. Hence, major advances in food safety control systems within developing countries will likely be a slow process without outside economic driving forces providing the resources needed to make and sustain change.

Private enterprise is one outside economic driving force that has been successfully incorporated into food production and processing operations of developing countries. Four different supply chain business models for the fisheries and aquaculture industry were described by Roth and Rosenthal (2006) and varied in the level of involvement of private enterprise. Complete involvement at all phases of the food continuum is one model and would include implementing food controls throughout production, processing, and transportation operations. This model gives the greatest level of control and safety. At the other end of the spectrum where the least amount of control over safety of imported foods occurs, the private enterprise selects their supplier but must monitor the safety of the incoming ingredients or product on a routine basis.

The latter model described above is essentially the system used by the FDA to regulate foods imported into the United States. Unfortunately, testing at the border may work for limited quantities of food, but as volumes escalate, as they have for the United States, the cost effectiveness of the task decreases considerably compared with other programs that could be implemented. Hence, the FDA has recently reevaluated its modus operandi and drafted the Food Protection Plan, an integrated strategy for protecting the nation's food supply (FDA, 2007b). This plan is characterized as a forward-oriented concept that uses science and modern information technology to identify potential hazards ahead of time. With an increased emphasis on prevention as the core of FDA's Protection Plan, several key recommendations would require legislative action and are listed below (FDA, 2007b).

1. Allow the FDA to require preventive controls to prevent intentional adulteration by terrorists or criminals at points of high vulnerability in the food chain.
2. Authorize FDA to issue preventive controls for high-risk foods.
3. Require food facilities to renew their FDA registrations every two years, and allow the FDA to modify the registration categories.
4. Authorize the FDA to accredit highly qualified third parties for voluntary food inspections.
5. Require new reinspection fees from facilities that fail to meet current Good Manufacturing Practices (cGMPs).
6. Authorize the FDA to require electronic import certificates for shipments of designated high-risk products.
7. Require new food and animal feed export certification fees to improve the ability of U.S. firms to export their products.
8. Provide parity between domestic and imported foods if FDA inspection access is delayed, limited, or denied.
9. Empower the FDA to issue a mandatory recall of food products when voluntary recalls are not effective.
10. Give the FDA enhanced access to food records during emergencies.

Recommendations 1 to 3 are elements that would assist in preventing foodborne contamination. Recommendations 4 to 8 are elements of intervention that seek to verify that the preventive measures are in fact being implemented in the correct manner. Recommendations 9 and 10 are elements that bolster FDA's emergency response efforts by allowing for increased speed and efficiency. Taken together, these ten actions would give the Agency the ability to collect information on the conditions under which the 189,000 registered foreign facilities (FDA, 2007b) manufacture, process, package, or store food exported to the United States as well as give them the tools to ensure compliance by these foreign facilities to written safety standards.

Currently, reliance by FDA on border inspections based on detection of hazardous contaminants is flawed. There are several routes by which pathogens can be introduced into foods; however, when contamination occurs, the harmful microbes are not distributed uniformly throughout the food. Under these circumstances, sampling programs and microbiological analyses cannot ensure their absence. Hence, negative results obtained through testing create a false sense of security. Implementation of pathogen prevention programs provides a system of programs that targets the source of contamination. Such programs do not replace verification testing programs, but they provide increased assurance of the safety of foods.

Food imported into the United States is increasing at an unprecedented rate, especially for horticultural products. The current emphasis on conversion of corn and other crops into ethanol for use as biofuels is likely to stimulate greater importation of foods. Many imported foods are produced in developing countries under conditions that can result in contamination by microbial pathogens. Fish and shellfish grown in ponds fertilized with untreated livestock or poultry manure or human feces, and fresh vegetables and fruits grown in night soil or with water containing human or livestock feces are examples of insanitary agricultural practices used in many countries that can lead to pathogen contamination. FDA testing has verified that contamination of such foods occurs; however, less than 1% of imported foods are laboratory tested and many of the sampling and testing procedures used are out of date. It is the food industry's responsibility to produce safe foods, which includes preventing or eliminating pathogen contamination. The government's role in ensuring the safety of imported foods is to verify the food industry is producing safe foods. Currently, there are major weaknesses in the food safety net for imported foods that need to be rectified by both the food industry and federal regulatory agencies for American consumers to be confident in the safety of the U.S. food supply.

REFERENCES

Ababouch, L. 2006. Assuring fish safety and quality in international fish trade. *Mar. Pollut. Bull.* **53**:561–568.

Allshouse, J., J. C. Buzby, D. Harvey, and D. Horn. 2003. International trade and seafood safety. USDA ERS Publication AER-828. [Online.] http://www.ers.usda.gov/publications/aer828/aer828i.pdf. Accessed 10 December 2007.

Asian Development Bank (ADB). 2004. Evaluation of Rural Water Supply and Sanitation in Indonesia. [Online.] http://www.adb.org/Documents/PPARs/INO/ppa-ino-26314.asp. Accessed 12 December 2007.

Becker, G. S. 2007. Food and agricultural imports from China. CRS Report for Congress. Updated 17 July 2007. [Online.] http://www.fas.org/sgp/crs/row/RL34080.pdf. Accessed 10 December 2007.

Bhandare, S. G., A. T. Sherikar, A. M. Paturkar, V. S. Waskar, and R. J. Zende. 2007. A comparison of microbial contamination on sheep/goat carcasses in a modern Indian abattoir and traditional meat shops. *Food Control* **18**:854–858.

Curtis, V., B. Kanki, S. Cousens, I. Diallo, A. Kpozehouen, M. Sangare, and M. Nikiema. 2001. Evidence of behavior change following a hygiene promotion programme in Burkino Faso. *Bull. W.H.O.* **79**:518–527.

Gran, H. M., A. Wetlesen, A. N. Mutukumira, G. Rukure, and J. A. Narvhus. 2003. Occurrence of pathogenic bacteria in raw milk, cultured pasteurized milk and naturally sourced milk produced at small-scale dairies in Zimbabwe. *Food Control* 14:539–544.

Harvey, P., and B. Reed. 2004. *Rural Water Supply in Africa. Building Blocks for Handpump Sustainability.* Water, Engineering and Development Centre, Loughborough University, United Kingdom.

Henson, S. 2003. The economics of food safety in developing countries. ESA Working Paper no. 03-19. [Online.] ftp://ftp.fao.org/docrep/fao/007/ae052e/ae052e00.pdf. Accessed 27 November 2007.

Huang, S., and F. Gale. 2006. China's rising fruit and vegetable exports challenge U.S. industries. USDA ERS Report FTS-320-01. [Online.] http://www.ers.usda.gov/Publications/FTS/2006/02Feb/FTS32001/fts32001.pdf. Accessed 7 December 2007.

Lopez, C. E., L. L. Ramos, S. S. Ramadan, and L. C. Bulacio. 2003. Presence of aflatoxin M_1 in milk for human consumption in Argentina. *Food Control* 14:31–34.

Montgomery, M. A., and M. Elimelech. 2007. Water and sanitation in developing countries: including health in the equation. *Environ. Sci. Technol.* 41:17–24.

Murray, C. J. L., and A. D. Lopez. 1996. *The Global Burden of Disease.* Harvard University Press, Cambridge, MA.

Roth, E., and H. Rosenthal. 2006. Fisheries and aquaculture industries involvement to control product health and quality safety to satisfy consumer-driven objectives on retail markets in Europe. *Mar. Pollut. Bull.* 53:599–605.

Salleh, N. A., G. Rusul, Z. Hassan, A. Reezal, S. H. Isa, M. Nichibuchi, and S. Radu. 2003. Incidence of *Salmonella* spp. in raw vegetables in Selangor, Malaysia. *Food Control* 14:475–479.

Unnevehr, L. J. 2000. Food safety issues and fresh food product exports from LDCs. *Agric. Econ.* 23:231–240.

U.S. Department of Labor. 2006. International comparisons of hourly compensation costs for production workers in manufacturing. [Online.] http://www.bls.gov/news.release/ichcc.t01.htm. Accessed 7 December 2007.

U.S. Food and Drug Administration (FDA). 2007a. FDA science and mission at risk. Report of the Subcommittee on Science and Technology, Appendix B. [Online.] http://www.fda.gov/ohrms/dockets/ac/07/briefing/2007-4329b_02_02_FDA%20Report%20Appendices%20A-K.pdf. Accessed 29 January 2008.

U.S. Food and Drug Administration (FDA). 2007b. Food Protection Plan. An integrated strategy for protecting the nation's food supply. [Online.] http://www.fda.gov/oc/initiatives/advance/food/plan.pdf. Accessed 12 December 2007.

Vij, V., E. Ailes, C. Wolyniak, F. J. Angulo, and K. C. Klontz. 2006. Recalls of spices due to bacterial contamination monitored by the U.S. Food and Drug Administration: the predominance of salmonellae. *J. Food Prot.* 69:233–237.

Index